建筑与都市系列丛书
Architecture and Urbanism Series

文筑国际 编译
Edited by CA-GROUP

Architecture and Beyond
Unconventional Practice by Chinese Architects

不止建筑
中国青年建筑师的跨界实践

中国建筑工业出版社

图书在版编目（CIP）数据

不止建筑：中国青年建筑师的跨界实践 = Architecture and Beyond Unconventional Practice by Chinese Architects / 文筑国际 CA-GROUP 编译 . -- 北京：中国建筑工业出版社 , 2024. 7. --（建筑与都市系列丛书）. -- ISBN 978-7-112-30092-1

Ⅰ . TU

中国国家版本馆 CIP 数据核字第 2024ZQ1211 号

责任编辑：毕凤鸣
责任校对：姜小莲

建筑与都市系列丛书
Architecture and Urbanism Series
不止建筑
Architecture and Beyond
中国青年建筑师的跨界实践
Unconventional Practice by Chinese Architects
文筑国际　编译
Edited by CA-GROUP

*

中国建筑工业出版社出版、发行（北京海淀三里河路 9 号）
各地新华书店、建筑书店经销
文筑国际制版
北京雅昌艺术印刷有限公司印刷

*

开本：880 毫米 ×1230 毫米　1/16　印张：16　字数：479 千字
2024 年 7 月第一版　　2024 年 7 月第一次印刷
定价：188.00 元
ISBN 978-7-112- 30092-1
　　（43509）
版权所有　翻印必究
如有内容及印装质量问题，请联系本社读者服务中心退换
电话：（010）58337283　QQ：2885381756
（地址：北京海淀三里河路 9 号中国建筑工业出版社 604 室　邮政编码 100037）

a+u

建筑与都市系列丛书
Architecture and Urbanism Series

总策划 Production
国际建筑联盟 IAM　文筑国际 CA-GROUP

出品人 Publisher
马卫东　MA Weidong

总策划人/总监制 Executive Producer
马卫东　MA Weidong　王　飞　WANG Fei

客座主编 Guest Editor
王　飞　WANG Fei

内容担当 Editor in Charge
完　颖　WAN Ying　吴瑞香　WU Ruixiang

英文校对 English Editing
唐小春　TANG Xiaochun

对话整理与翻译 Conversation editing and translation
左　赢　ZUO Ying　　　齐聿铭　QI Yuming　　李硕烜　LI Shuoxuan
罗辰浩　LUO Chenhao　齐聿美　QI Yumei　　　李　睿　LI Rui
朱海珲　ZHU Haihui　　张雅淇　ZHANG Yaqi　　庄宇锟　ZHUANG Yukun
薛皓天　XUE Haotian　　孟昕琦　MENG Xinqi　　骆霈语　LUO Peiyu

封面图 Front cover
孙福瑞　SUN Furui

封底图 Back cover
绘造社　Drawing Architecture Studio
雪城大学建筑学院　Syracuse University School of Architecture

书籍设计 Book Design
文筑国际　CA-GROUP

中日邦交正常化50周年纪念项目
The 50th Anniversary of the Normalization of
China-Japan Diplomatic Relations

A+U Publishing Co., Ltd.
发行人/主编：吉田信之
副主编：横山圭
编辑：服部真吏　Sylvia Chen

本系列丛书著作权归属文筑国际，未经允许不得转载。本书授权
中国建筑工业出版社出版、发行

A+U Publishing Co., Ltd.
Publisher / Chief Editor: Nobuyuki Yoshida
Senior Editor: Kei Yokoyama
Editorial Staff: Mari Hattori Sylvia Chen

The copyright of this series is owned by CA-GROUP. No reproduction without permission. This book is authorized to be published and distributed by China Architecture & Building Press.

Preface
The Blur of Practice

Michael Speaks

This book does not concern itself—as have so many "timely" architecture publications over the last quarter-century—with the presentation of a new architecture movement intent on leading the way into the future. What is presented here is instead a version of the future of architecture and architecture practice that has already arrived in China. As such, it can be viewed as a "retroactive manifesto" that offers lessons gleaned from the future-past. Before turning directly to those lessons, and to provide some context for what follows, it is important first to review some recent history.

The 1990's saw the rise of a Dutch avant-garde proclaimed as heralds of a new era of global modernization. Led by Rem Koolhaas and OMA, whose monograph, *S,M,L,XL* (1995), served as diary, travelogue and how-to manual, the Superdutch and architecture avant-garde competitors flying the flags of Deconstructivism, Parametricism, Minimalism, and others, led the way onto an emergent continent of opportunity in East Asia, especially in China. Many of these firms took full advantage of their vanguard status and completed, throughout the region, a host of exhilarating new projects. But with the onset of the 2008 economic crisis, which commenced a global economic unraveling that is still underway, it became clear that the Superdutch and their ilk were not vanguard heralds of a new era but were instead harbingers signaling the end of an old one birthed in Europe and grown to maturity in America. It thus became clear in 2008, as the world watched on Chinese-made television sets an Olympic Games broadcast from Beijing and staged in buildings designed by these same Western architecture avant-gardes, that the future of global modernization would be decided in East Asia and China, not in Europe and not in America.

Anyone who has been to Shenzhen will know that in China modernization is a churning aggregation that everywhere, from the urban village to the stock exchange, moves at different speeds—in some precincts at a snail's pace and in others at hyperspeed, at what locals call Shenzhen speed. Not only does the hypermodern sit cheek-to-jowl with the pre-modern, but also it is larded into the fabric of the modern and pre-modern, rendering meaningless normal spatial and temporal lines of demarcation between past, present and future. Beginning in the 1990s and quickening each year since, this temporally and spatially marbled aggregation came to absorb and ultimately supersede the linear, Western manifestation of modernization that many believed would colonize all space and time, reducing local, regional and national difference into a uniform global sameness. Driven by speed, by the fast and the slow, this Chinese manifestation of global modernization rendered obsolete the architecture avant-garde's ambition to chart a future course for all to follow. What is the future, after all, if, as one science fiction writer famously observed, it is already here, just not evenly distributed? During this same period an alternative set of ambitions and practices with new attributes—agility, collaboration, professional shapeshifting, to name but a few—better suited to the demands of speed emerged in China and are, along with the designers that embody them, the primary focus of this book.

All those included in the pages that follow were trained as architects and all practice in China, though none practice architecture in any conventional sense. Many, including PENG Wu (Glodon) are researchers concerned with the production of new knowledge. Others, including SU Qi (Modelo), QIU Wenhao (iFLYTEK), LUO Feng (Onesight Technology), Leslie Lok (HAHHAH) and DING Junfeng (FABO) are makers and technologists who focus on new building methods, systems and materials. While some of the designers could be labeled "slashers," a term used to describe creatives in China with multiple professional identities, their creative practices are enabled rather than defined by their disciplinary training as architects. The backslashes (//) and the professional identities defined between them (//technologist // architect // car designer//) are thus less important than the blur that occurs as each, according to the demands of the project and guided by their own creative responses to those demands, becomes something other than an architect: a film maker, or fashion designer, fabricator, technologist, or researcher—each blur itself a creative act rather than a fixed identity worn like a mask by the architect.

When FAN Ling (Tezign) creates and operates a successful AI firm, and MENG Hao (Robotic Plus) launches a robotics construction company, they are not architects wearing a mask or playing the role of CEO, but are instead design entrepreneurs who use their skills and training to become CEOs in the same way that MA Yuanrong (GEIJOENG) uses her skills to become an Art Director, then a Yoga Instructor; JU Bin (Horizontal Design) uses his skills to become a yacht designer and then a product designer; and WANG Zhenfei and WANG Luming (FUN Design) use their skills to become furniture designers and then researchers—all in response to the ever changing demands of global modernization in China. Like WANG Zigeng (PILLS), who has designed a series of extraordinary installations and "film sets," made a series of films, and taught at the famed Beijing Film Academy, these designers are empowered by, but not limited to, the architecture discipline. In this way they all hew more closely to the contours of practice shaped by architect-trained designer and cultural influencer Virgil Abloh than to avant-garde practices of architecture like OMA, led by architect Rem Koolhaas or BIG, led by architect Bjarke Ingels, the last of the Superdutch, a designation conferred not by birthright but earned by transforming innovation—the watchword and most

序言
实践的模糊性

迈克尔·斯皮克斯

本书并不像过去二十五年来的许多"及时"建筑出版物那样，致力于展示一个引领建筑学未来的新运动。这里呈现的是一个早已在中国发展的未来建筑学及其实践。因此，本书将作为一部追溯的作品，从过去汲取对未来的启发。在直接进入本书的内容以及铺垫相关的背景信息之前，回顾一下三十余年的历史是非常重要的。

20世纪90年代，荷兰先锋派的兴起如同先驱一般，开创了全球现代化的新时代。由雷姆·库哈斯及其大都会建筑事务所编写的著作《S，M，L，XL》（1995），可视为一本日记、游记和指导操作手册，超级荷兰和建筑先锋派高举着参数化和解构主义的旗帜，引领东亚（特别是中国）进入一个新兴的、相互关联的、充满机遇的全球化世界。许多公司充分利用了这一优势，在整个地区完成了一系列令人振奋的新项目。但随着2008年经济危机的爆发，全球经济随之陷入长期的低迷。很明显，荷兰的超级公司们并没有拉开新时代的序幕，反而预示着这个开始于欧洲、繁盛于美国的旧时代的终结。因此，2008年，当全世界通过中国制造的电视机观看从北京转播的在这些前卫建筑中上演的奥运会赛事时，这一点变得清晰起来：全球现代化的未来将由中国决定，既非欧洲亦非美国。

去过深圳的人都知道，中国的现代化是一个剧烈运转、不断整合的过程。从城中村到证券交易所，所到之处无不以各自的速度前进着——其中有些地区发展速度如同蜗牛漫步，但有些地区的发展速度快到令世人瞠目，就像当地人常说的"深圳速度"。超现代不仅承接了现代化初期的发展，而且极好地融入近代现代化的框架中，使过去、现在和未来之间的三维时空融为一体。从20世纪90年代开始，现代化进程每一年都在加速，这种时间和空间上如同磁铁石般多元自由的发展模式，会吸收并最终取代西方现代化的线性过程。许多人担心这样的现代化将统治所有的空间和时间，减少地方、区域和国家的差异性，最后变得全球统一，但中国的发展模式取决于发展速度的快与慢，而非空间和时间，中国在全球现代化进程中的表现让建筑先锋派体恤全人类的雄心变得过时。最后，未来是什么？正如小说家威廉·吉布森（William Gibson）的名言所说，未来已经存在，只是分布得不均匀。同时期诞生的理想、实践以及衍生出的新的品质——敏捷、团队协作、专业能力以及跨界转型——都能更好地满足中国和其他地方出现的关于速度的需求。本书就将聚焦具有这样精神的年轻设计师们。

接下来提到的所有设计师虽然都接受过建筑专业的培训，并且都在中国从事建筑实践，但他们并没有从事传统意义上的建筑设计实践。有许多人，包括何勇（好处）、彭武（广联达）、杨小荻（小库科技）和冯果川（童筑）都是关注新知识生产的研究人员，其他人则是专注研发新建筑技术的创客，如苏奇（模袋）、邱文浩（科大讯飞）、罗锋（以见科技）、陆唯佳（HAHHAH）和丁峻峰（FABO）。本书中的一些设计师可能被贴上"斜杠青年"的标签，这是一个用来形容具有多种专业或学科背景的中国创意人士的术语，他们的创作以及实践是基于他们所受到的建筑学教育，但却并不被其束缚。事实上，正是他们所受的训练和对于专业的敏感，让他们能够在深圳、上海和北京成长起来。因此，"斜杠"符号（//）与它所定义的字面身份（// 建筑师 //）已不再重要，重点在于其背后所代表的领域与专业的糅合。每位设计师根据项目的需求和他们自己对这些需求的回应，成为电影制作人、时装设计师、制造商或技术专家，是不断尝试的创意与创新让他们的身份变得神秘，他们绝非为了掩盖过时的专业身份而戴上面具的建筑师。

当范凌（特赞）创建并运营一家成功的人工智能公司，孟浩（大界机器人）推出一家机器人建筑公司时，他们不是扮演CEO角色的建筑师，而是利用自己的技能和实践的经验成为CEO的设计企业家。就像马圆融（幾樣）成为艺术总监和瑜伽教练一样，琚宾（水平线设计）成为游艇设计师，王振飞和王鹿鸣（FUN）成为工业设计师，张烁（烁设计）成为平面设计师——这一切都是为了响应中国在全球现代化中

important feature of all 90s Dutch firms—into an avant-garde ideology with a defined formal language.

Abloh, a Black designer born near Chicago after his family moved there from Ghana, was trained as an architect, worked as a designer and creative director at Kanye West's Donda Creative Agency, and founded his own fashion label, Off-White, before being tapped in 2018 to become the men's artistic director for fashion powerhouse Louis Vuitton. At the time of his tragic death at age 41 in November 2021, Abloh was among the most important designers and cultural influencers in the world. Abloh was a generational, global design talent whose legacy will be defined more by the ways he shaped design practice than by any design he realized for the many companies he worked with including Nike, Alessi and Ikea, to name only a few. Abloh reset the ambitions for a new generation of architects and designers in two important ways. First, he disavowed the avant-garde ambition to "make it new," embracing instead the Duchampian concept of the ready-made. With his famous 3% approach, Abloh argued that it was impossible and unnecessary to invent something entirely new and that the designer need only edit a found object by 3% to transform and differentiate it from what already exists. He retrieved the "already there" from the dustbin of history, to which the avant-garde had consigned it, and used it as the material for all "new" design. Second, Abloh was less interested in the design object itself and more in how any design—a sneaker, a chair or a showroom—could become a vehicle for communication and audience building. He used Instagram, WhatsApp and other social media platforms the way architects use pencils, mouses and sketch pads. But while architects use their tools to design buildings, Abloh used his to transform design objects and experiences into a social adhesive that binds together audiences, communities, and collaborators—his intended design object and ultimate design objective.

TANG Yu (Archmixing), LI Jingjun (Li Lexian Media), QIN Qingxia (Zidao Culture), YIN Yujun (AAA), and ZHANG Shuo (SURE Design) all embrace this latter form of design practice, and like Abloh each has made design more accessible, interactive and more appealing to young design audiences. TANG Yu is the producer of numerous successful videos focused on contemporary architecture in China and, beginning in 2016 with the founding of an offshoot company, Nextmixing, he expanded the size and profile of the firm's audience through the development of new platforms that facilitate the planning and execution of design related events, forums and conferences. LI Jingjun and QIN Qingxia are among the most important and prolific KOLs (Key Online Leaders), in China, and their massive online following provides them with the kind of platform that avant-garde magazines and journals from the last century could never have imagined. QIN Qingxia, for example, has created through his Zidao Culture platform an unimaginably large audience for "The Furious Yama," among the most widely consumed manga in China. YIN Yujun is also a designer of architecture and design audiences, though his medium is the digital-analog format evident in his pioneering and long-standing curatorial work at the Bi-City Biennale of Urbanism\ Architecture in Shenzhen. YIN has often collaborated on these projects with ZHANG Shuo, a brilliant visual communicator who uses a variety of techniques and media to tell stories. Those skills are evident in the comic series included in this book where he makes use of simple lines and drawings to stretch and warp the geometry of conventional spaces, creating an alternative reality that rivals any video game, VR or augmented reality environment. ZHANG's collaboration with FENG Guochuan (Archild) on a series of children's projects intended to raise their awareness of architecture and design, creates the most important audience of all: future architects and designers.

While they do not subscribe to Abloh's 3% approach, LI Han and HU Yan (Drawing Architecture Studio), WANG Xin (Gardening Architecture), and ZENG Renzhen (Yushanfankuan) do eschew the avant-garde obsession with "the new" and work within existing traditions. To meet the representational demands of the image-saturated hyperreality that is contemporary China, LI Han and HU Yan have developed a unique drawing method that they employ to transform pedestrian scenes of everyday urban life into magical, fictive universes of characters, stories and places. Reminiscent of the ambitions—but not the techniques—of Chinese landscape painting, they employ a variety of visual tools—photography, comic strips, axonometric drawing—to "document" the slow, and sometimes fast, pace of urban transformation, revealing details large and small that are often overlooked by those seeking only to see or invent "the new." Indeed, their drawings, perhaps like no other contemporary visual design medium, accedes to the ambitions of Chinese landscape painting and gardens that, as described by WANG Shu, are meant to serve as "recollective apparatuses and mythic machines," which reconfigure multiple experiences of a site, city or world. WANG Xin also works within established traditions of Chinese landscape garden design and its attendant literati language and philosophical concerns to slow the pace of urban life and create opportunities for reflection and contemplation. His interest in the future is connected to and contextualized by a deep understanding of the past, as he observes: "Tradition is also a component of the future, tradition is a repository idea for the future, a healing touch, and a brake for the future. Nowadays, in a society that is running wildly forward, we cannot only have a throttle, but also a brake." In "Tiger Arts" and "Little Cave Sky," he also reveals that applying the brake, slowing down, is not motivated by a naïve desire to return to a more simple, natural world, but instead by the studied intention to reconstruct, for our contemporary world, the sophisticated and mannered contemplation of man's relationship to the natural world evident in traditional paintings and garden design. ZENG Renzhen's paintings, many of which were originally intended as landscape studies, provide another unique and beautiful treatment of traditional and contemporary subjects, scenes and motifs. In the "Garden of Fantasies," he has, through a four-stage process, adapted and transformed for contemporary audiences many such scenes but also the techniques and ambitions of traditional Chinese painting. In the "Red Series" and the "Green Series," some of the same lessons can be observed but in a less analytic and more applied approach that reveals, as do the paintings of Zeng Renzhen, a highly mannered, sophisticated depiction of the relationship between man and nature.

This publication ultimately shows that today global modernization has the same differential shape, structure and form in Chicago, Paris and Tokyo, as it does in Beijing, a shape, structure and form that rewards practitioners who blur the professional lines that confine them so that they may better respond to a world shaped by speed rather than one delineated and bounded by space and time. It also makes clear that we all have a great deal to learn from China and from those whose work can be found in the pages that follow.

的需求。就像设计了一系列精彩装置和"电影布景"、制作了一系列电影并在著名的北京电影学院任教的王子耕（PILLS工作室）一样，这些设计师接受但不局限于建筑的学科边界。通过这种方式，他们都更接近于由受过建筑学教育的设计师和文化引领者维吉尔·阿布卢赫塑造的形象，而不是由建筑师雷姆·库哈斯领导的 OMA 或由建筑师比亚克·英厄尔斯领导的 BIG 等传统的跨学科建筑实践，也不是像超级荷兰一样，其称号并非与生俱来，而是将创新（所有 20 世纪 90 年代荷兰公司的口号和最重要特征）通过明确的建筑语言转变为具有明确辨识度的前卫意识形态。

阿布卢赫是一名黑人设计师，在全家从加纳来到美国后出生于芝加哥附近，接受过建筑学教育，在坎耶·欧马立·韦斯特的 Donda Creative Agency 担任设计师和创意总监，并在 2018 年被路易威登聘用成为男装艺术总监之前，创立了自己的时尚品牌 Off-White。阿布卢赫于 2021 年 11 月不幸去世，享年 41 岁，他是世界上最重要的设计师和文化影响者之一。阿布卢赫是一位世代相传的全球设计人才，他对于设计界的宝贵遗产更多取决于他影响和塑造设计实践的方式，而不仅仅是他为包括耐克、阿莱西和宜家在内的许多公司所设计的具体产品。阿布卢赫在两个重要方面重新设定了新一代建筑师和设计师的创意抱负。首先，他并不赞同先锋派"让它焕然一新"的野心，而是接受杜尚式的现成品概念。凭借他著名的 3% 方法，阿布卢赫坚持认为设计师不可能发明全新的东西，只需将已存在的物品编辑 3% 即可将其与现有事物进行转换和区分。其次，阿布卢赫较少关注设计本身，而更多地关注在将设计用作和客户沟通的手段。

唐煜（阿科米星）、李憬君（李乐贤传媒）、覃清硖（子道文化）和尹毓俊（AAA）都接受了后一种的设计实践形式。唐煜是聚焦于中国当代建筑的视频制作人，制作了许多非常成功的视频，并于 2016 年成立了一家分支公司那行文化（现更名为米行文化），他通过研发新平台来宣传公司形象并扩大公司的规模。这些新平台同时带动了设计相关的活动、论坛和会议的开展。李憬君和覃清硖是中国最重要和最多产的建筑意见领袖，650 万粉丝为他们提供了足以影响建筑和设计领域实践和潮流的平台，这种影响力是 20 世纪任何前卫杂志和期刊都无法想象的。尹毓俊也是建筑和设计领域的意见领袖和设计师，但他的作品是以混合的、数字模拟的形式呈现的，这在他深圳双年展中开展的开创性和长期性的策展工作中充分体现。以上所有设计师都让设计更容易和大众互动以及更好地被理解，也更能吸引新一代的观众群体。

尽管他们不完全契合阿布卢赫的 3% 理念，但李涵和胡妍（绘造社）、王欣（园林建筑）以及曾仁臻（鱼山饭宽）均避开了前卫对"新"的迷恋，而是在传统回归和日常材料的框架内展开各自的实践。为满足当代中国充斥着图像的超现实要求，李涵和胡妍独创了一种绘画方法，将日常城市生活中平凡的场景转变为充满角色、故事和场所的奇幻虚构世界。他们运用多种视觉工具，包括摄影、漫画连环画、轴测图，以"记录"城市转变的缓慢或快速步伐，揭示了那些常被仅寻求"新"事物的人所忽视的大小细节。事实上，他们的绘画，或许是当代其他视觉设计媒体所无法企及的，承袭了中国山水画和园林的宏大抱负，正如王澍所言，这些作品旨在充当"回忆装置和神话机器"，重构一个场地、城市或世界的多重体验。王欣同样坚守中国传统园林设计及其伴随的文人语言和哲学关切，以减缓城市生活的节奏，创造反思和冥想的机会。他对未来的兴趣紧密联系着对过去的深刻理解，正如他所言："传统也是未来的一部分，传统是未来的理念库，是一种治愈的触摸，是未来的刹车。如今，在一个疯狂前进的社会，我们不能仅有油门，还需要刹车。"在"虎艺术"和"小山洞天"中，他也展现了应用刹车、减速的追求并非出于幼稚的愿望回归到更为简单自然的世界，而是出于深思熟虑的意图，为当代世界重新构建传统绘画和园林设计中体现的人与自然关系的精妙而有仪态的冥思。曾仁臻的绘画，其中许多最初作为山水研究，提供了另一种独特而美丽的对传统和当代的主题、场景和图案的处理。在"幻园"中，他通过一个四阶段的过程，为当代观众改编和转化了许多这样的场景，同时也改编了传统中国绘画的技巧和抱负。在"红系列"和"绿系列"中，可以观察到一些相同的经验，但更注重应用，正如曾仁臻的绘画所展示的，高雅地、精致地描绘了人与自然之间的关系。

本书最终揭示的是，今天的全球现代化在芝加哥、巴黎、东京和北京有着完全不同的形态、结构和形式。这种形态、结构和形式需要从业者模糊学科和专业之间的界限，不被它们所限制，从而更好地回应一个由速度塑造的世界，而非一个由空间和时间描绘和限定的世界。我们都还需要向中国、向阿布卢赫以及后文提到的设计师们学习很多。

Michael Speaks

A writer, editor, curator, educator, and dean and professor at the Syracuse University School of Architecture

迈克尔·斯皮克斯

作家、编辑和教育家，美国雪城大学建筑学院院长及教授

Architecture and Beyond
Unconventional Practice by Chinese Architects

Preface
The Blur of Practice   6
Michael Speaks

Editor's Word   13
WANG Fei

Architect Index   14

Design and Beyond   16

Dialogue:
Design and Beyond   18
JU Bin / WANG Zhenfei / WANG Luming / WANG Zigeng / WANG Xin / MA Yuanrong / HE Yong / ZHANG Shuo

Horizontal Studio
Windrider 1   30

FUN Design
Grand Gourmet Flagship Store   42

PILLS
Nine-Tiered Pagoda PSFO Exhibition Space Design   50

PILLS
Nineteen Ninety-Four   58

Garden Architecture
Little Cave-Sky   66

Garden Architecture
Asia Bamboo Life and Art Exhibition   74

SURE Design
Beyond Graphic   82

Technology and Beyond   88

Dialogue:
Technology and Beyond   90
FAN Ling / MENG Hao / PENG Wu / SU Qi / QIU Wenhao

RoboticPlus.AI
Intelligent Construction—Future of Architecture   102

HANNAH
House of Cores   120

HANNAH
Ashen Cabin   126

Onesight Technology
Qingdao Ruyi Lake Complex   134

FABO
Casablanca Biennale —"Chopsticks"   138

Essay:
Software Driven Robotics for Intelligent Construction   148
LIANG Zee / LAI Kuan-ting / MENG Hao

Essay:
Architect and Architecting: The Third Wave of Architecture in the Age of New Technology   160
FAN Ling / CHEN Xueer

Media and Beyond   170

Dialogue:
Media and Beyond   172
LI Han / HU Yan / ZENG Renzhen / YIN Yujun / TANG Yu / FENG Guochuan / QIN Qingxia / LI Jingjun

Drawing Architecture Studio
The Complete Map of Capital Beijing   184

Drawing Architecture Studio
The Grand Stage   190

Yushanfankuan
Garden of Fantasies   196

Atelier Alternative Architecture
Spatial instigation: village, factory, city   204

NEXTMIXING
NEXTMIXING- The Spatial Experiments on Usage   226

Archild
Archild   238

Zidao Culture
Shituzi Architecture Channel   242

Interview:
After Architecture   244
Aric CHEN

Profile   252

不止建筑
中国青年建筑师的跨界实践

序言
实践的模糊性　7
迈克尔·斯皮克斯

编者的话　13
王飞

建筑师索引　14

不止设计　16

对话：
不止设计　18
琚宾 / 王振飞 / 王鹿鸣 / 王子耕 / 王欣 / 马圆融 / 何勇 / 张烁

水平线设计
御风者1号　30

凡语设计
Grand Gourmet 旗舰店　42

PILLS 工作室
"九层塔"政纯办个展展览空间设计　50

PILLS 工作室
1994年　58

造园建筑
小洞天　66

造园建筑
"东方竹"亚洲竹生活艺术展序幕空间设计　74

烁设计
超越平面　82

不止科技　88

对话：
不止科技　90
范凌 / 孟浩 / 彭武 / 苏奇 / 邱文浩

大界机器人
智能建造——建筑的未来　102

HANNAH
核心之屋　120

HANNAH
梣木小屋　126

以见科技
青岛如意湖　134

"数制"工坊
卡萨布兰卡艺术双年展——"筷子"　138

论文：
面向智能建造的新型机器人技术　148
梁喆 / 赖冠廷 / 孟浩

论文：
建筑师和架构：新技术可能下的第三浪建筑学　160
范凌 / 陈雪儿

不止媒介　170

对话：
不止媒介　172
李涵 / 胡妍 / 曾仁臻 / 尹毓俊 / 唐煜 / 冯果川 / 覃清硕 / 李惜君

绘造社
京师全图　184

绘造社
大戏台　190

鱼山饭宽
幻园　196

多样建筑
空间策动：村，厂，市（以研究与展览作为空间策动的工具）　204

那行
那行——使用的空间实验　226

童筑文化
童筑文化　238

子道文化
使徒子建筑频道　242

访谈：
建筑之后　244
陈伯康

设计师简介　252

Editor's Word

编者的话

Since the onset of the 21st century, the world has undergone profound transformations encompassing the realms of politics, economics, society, culture, and the environment. This evolution has not left the field of architecture untouched, presenting young architects in China with challenging decisions as they navigate this shifting landscape. China has witnessed rapid urban development in the decades following its opening-up, but as urbanization slows, the architecture industry is undergoing a process of refinement. This transformation is ushering in a new breed of unconventional architects in China, who are pushing the boundaries of conventional design and championing diversity in development, design, management, and the industry as a whole. They are seamlessly blending technology and industry to propel architecture and design to unprecedented heights, scopes, and depths. These architects are not limited to traditional media but are also venturing into new media, forging a fresh synergy between architecture and media. In this book, we have enlisted the insights of architectural critic and educator, Michael Speaks, to pen an essay titled "The Blur of Practice," redefining the contemporary architect's identity. Additionally, three partners from RoboticPlus.AI delve into the world of software-driven robotics for intelligent construction. Meanwhile, technology entrepreneur and architecture scholar, FAN Ling, examines the profound impact of technology on architecture. Design and art curator, Aric Chen, explores architecture's far-reaching influence on other domains. Through dialogues with unconventional architect professionals on subjects spanning design, technology, and media, we shed light on the multifaceted nature of architecture and its influence that transcends traditional boundaries. These pioneering projects showcased herein represent the emerging generation of unorthodox architects in China.

(a+u)
(WANG Fei)

二十一世纪以来，世界经历了天翻地覆的变化，政治、经济、社会、文化、环境无一不经历着巨变，建筑亦然。新一代的建筑师面临着重大的抉择，特别是中国的新一代建筑师。改革开放之后，中国经历了几十年迅猛的城市发展，近年来发展速度放慢，建筑产业也更为细化，非常明显地涌现出越来越多非传统的中国建筑师。他们挑战和突破传统设计的边界，将开发、设计、运营、产业更为多元地推进；他们将科技与产业结合，推动建筑产业走向新高度、广度与深度；他们既专注传统媒体也关注新媒体，将建筑与媒介挖掘出更多的维度。本书邀请了著名建筑理论家及教育家迈克尔·斯皮克斯撰文"实践的模糊性"，对当代建筑师的新身份进行重新定义；大界机器人的三位合伙人展开论述了智能建造的新型机器人技术；科技创业家建筑博士范凌教授讨论科技对建筑和城市的影响；与著名设计与艺术策展人陈伯康对话讨论建筑学对其他身份的影响；与二十余位非建筑的建筑师的讨论从设计、技术和媒介的角度展示不止建筑的多元性；通过多个非传统的项目和作品向大家介绍中国新一代非传统建筑师的当下。

(a+u)
（王飞）

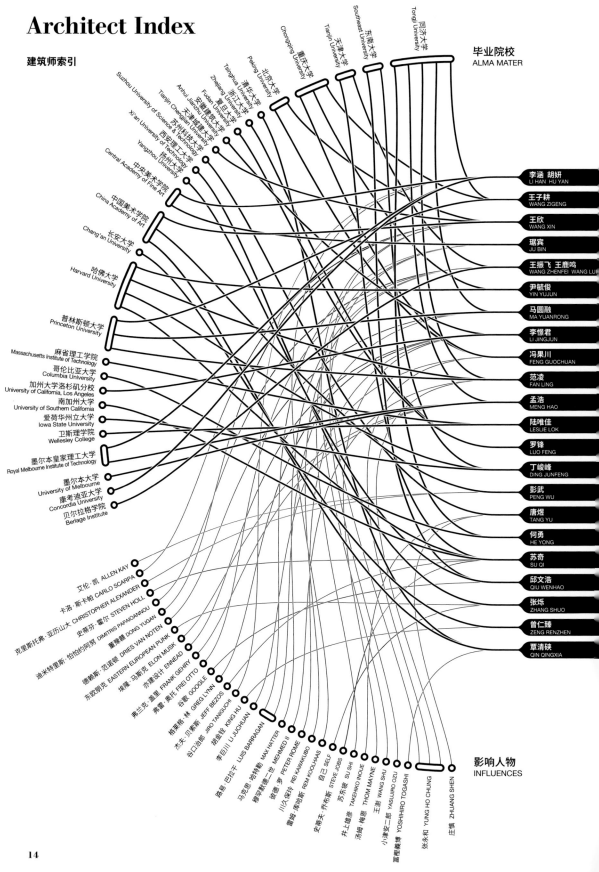

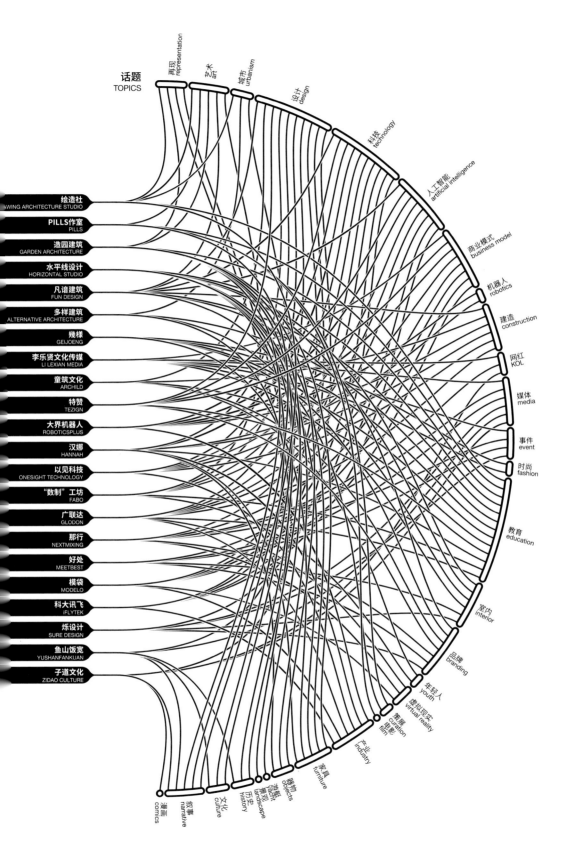

Design and Beyond
不止设计

王子耕 WANG ZIGENG
王欣 WANG XIN
琚宾 JU BIN
王振飞 王鹿鸣 WANG ZHENFEI WANG LUMING
张烁 ZHANG SHUO

Windrider 1 御风者 1 号
Horizontal Studio 水平线设计 (pp. 30–41)

Grand Gourmet Flagship Store Grand Gourmet 旗舰店
FUN 凡谙设计 (pp. 42–49)

Nine-Tiered Pagoda PSFO Exhibition Space Design
"九层塔"纯政办个展览空间设计
PILLS 工作室 (pp. 50–57)

Nineteen Ninety-Four
PILLS 工作室 (pp. 58–65)

Little Cave-Sky 小洞天
Garden Architecture 造园建筑 (pp. 66–73)

Asia Bamboo Life and Art Exhibition "东方竹"亚洲竹生活艺术展序幕空间设计
Garden Architecture 造园建筑 (pp. 74–81)

Beyond Graphic 超越平面
SURE Design 烁设计 (pp. 82–87)

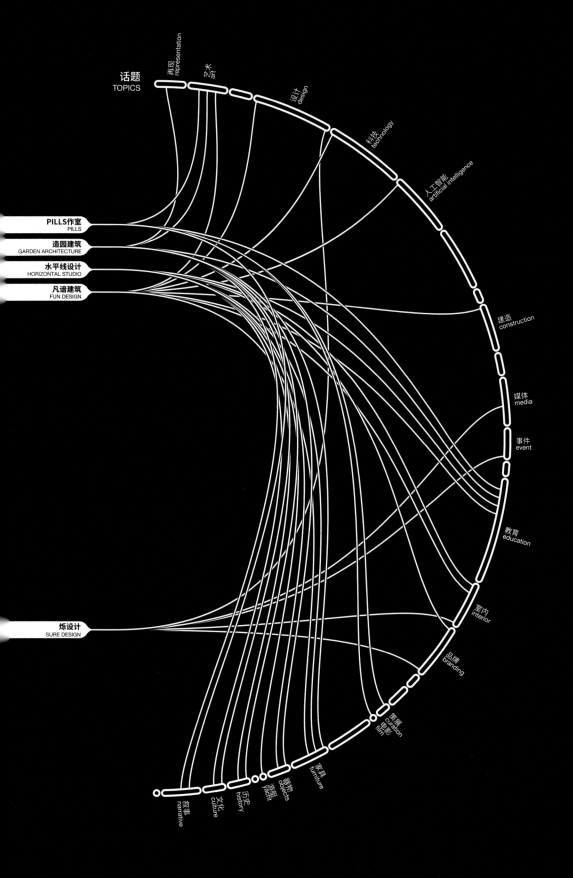

Dialogue:
Design and Beyond
JU Bin / WANG Zhenfei / WANG Luming / WANG Zigeng / WANG Xin / MA Yuanrong / HE Yong / ZHANG Shuo

对话:
不止设计
琚宾 / 王振飞 / 王鹿鸣 / 王子耕 / 王欣 / 马圆融 / 何勇 / 张烁

WANG Fei: Today, we engaged in conversations with seven unconventional architects, all of whom have a background or current involvement in architectural design, as well as multiple design identities such as interior designer, fashion designer, product designer, graphic designer, lifestyle designer, space designer, film scene designer, etc. I would like to start a discussion about your definition of "design" and how architecture has influenced your other identities.

MA Yuanrong (GEIJOENG), as a registered architect in the USA, a professional yoga instructor, model, and fashion designer, what is the relationship between architecture and your other identities?

MA Yuanrong: Currently, I am mainly involved in art guidance within the fashion industry, including spatial and interior design. In fact, this is not too far from architecture, although there may be differences in terms of project scale and timeframes compared to traditional professional architects. There are many interesting contrasts and differences between architecture and fashion, and in some ways, they can be considered opposite poles. They also involve some discussions about aesthetic gender to some extent, which were covered in the last semester. Architecture is generally perceived to be dominated by male aesthetics - solid, rigid, static, and eternal. In contrast, the fashion industry is oriented towards female aesthetics - soft, flexible, versatile, and fleeting. Therefore, practicing in both fields at the same time can generate doubts and reflections, provide additional perspectives on issues, and improve sensory acuity.

The practice of yoga has made me pay more attention to the connection between the body and space - how the energy field emanating from a place can affect an individual's energy, and how the space atmosphere can regulate emotions, etc. Additionally, there are some dualistic dialectics and inspirations from the theory of non-duality, such as the yin and yang of the human body, inner and outer, strength and flexibility, and the aforementioned architecture and fashion - the unity of opposites between male and female aesthetics, as well as the relationship between rationality and irrationality, emotion and thought, etc. When these intersect, the way of dealing with conflicts when designing will be different.

WANG Fei: WANG Zhenfei and WANG Luming (FUN), your design projects use mathematics to connect, experiment, contract, and extend through mathematical scales and types. Do you consider mathematics as the highest hierarchy, or the core value, or something else?

WANG Luming: In my opinion, an architect's work can be divided into two major parts. The first part involves the manifestation of the work itself and representation of social responsibilities. The second part is more about individual preferences that manifest in the work. Social responsibility is a broad topic, but its implementation is restricted by many uncontrollable factors. I can express my understanding of the world more easily through individualistic preferences and visions. Mathematics is a part of our interest and world

王飞： 今天我们与七组非常规的建筑师进行对话，大家都曾经或者正在从事建筑实践，大家目前也都有着多重设计师的身份，建筑师、室内设计师、服装设计师、产品设计师、平面设计师、生活方式设计师、空间设计师、电影场景设计师等，我想就大家对"设计"的定义以及建筑学对你的影响进行展开讨论。

马圆融（幾樣），作为一位美国注册建筑师、职业瑜伽教练，也是一位模特、时装设计师，建筑学与你其他身份的关系是怎样的？

马圆融： 目前的话，主要是在从事时尚行业的一些艺术指导，包括一些空间和室内设计，其实跟建筑的关系还不算太远，但可能同传统意义上的职业建筑师在项目尺度和时间性上有些差别。建筑和时尚两个领域存在很多有趣的对立和反差，在某些维度几乎是完全相反的两极，也在一定程度上牵涉一些关于审美性别化的讨论，这点我们在上学期的课程中也有涉及。建筑学一般被人所认知的是以男性审美为主导的，坚固、刚硬、静止、追求永恒，而时尚行业则以女性审美为导向，柔软、弹性、多变、稍纵即逝，所以同时在这两个领域进行实践，会比较容易产生一些怀疑和反思，增加一些看问题的视角，也会提升一些感官上的敏锐度。

瑜伽的习练让我更加关注到身体和空间体的连结，场所散发的能量场能够如何影响个体的能量，空间氛围可以如何去调节个人情绪等等。此外还有一些二元辩证以及不二论的启发，比如人体的阴阳两脉、内部与外部、力量与柔韧和刚才提到的建筑与时尚、男性与女性审美上的对立统一，以及理性和非理性、情绪与思维等等的关系，这些关系交叉在一起，做设计的时候处理矛盾的方式就会不太一样。

王飞： 王振飞和王鹿鸣(FUN)，通过数学的"寻"规"导"矩

Geijoeng at MixC Qianhai, Shenzhen © Geijoeng
幾樣深圳前海万象城 © 幾樣

Appollonian Gasket 01, 2017 © FUN
阿布洛尼垫圈 1 号，2017 © FUN

view. My understanding of the world is based on a certain type of logic. Our way of thinking is formed by the fundamental education during our younger years, and we are a product of our generation. The architecture profession has various subjective opinions that differ from one architect to another. Because of the diversity behind the thought process of our profession, we hope to demonstrate the production process like that of artists. It is through this process that architects can obtain the freedom of individual expression.

WANG Zhenfei: I met my teacher, Peter Trummer, at the Berlage when I was studying there. During his studio, Peter explained to me that when the top point of a triangle moves on a line parallel to its opposite side, the triangle's shape changes without changing its area. At that time, I felt as though my left and right brain finally connected because of this simple equation. My dormant knowledge of mathematics was activated alongside my design skills. A lot of formal variations can be achieved without changing the area, and this finding can have real-world applications. I began to slowly find my way in design. This type of thinking can bridge mathematical thinking and design, and we can solve many architectural challenges through the careful implementation of mathematical thinking. I then started to research in this direction and ultimately employed this method for my thesis on the residences of the Southern Chinese water towns.

After returning to China and starting our own firm, I encountered many more opportunities for practical implementation. In the beginning, I worked on exhibitions, installations, and small-scale furniture. Because designing and implementing on a smaller scale took less time, I was able to efficiently verify my theories through faster iterations. It was important for me to learn from trials and errors and correct my mistakes. During these few years of practice, we researched and experimented with converting mathematical thinking into design thinking. However, the knowledge we used was scattered, and we struggled to systematize it. In 2015, *UED* magazine invited us to publish a special edition, which prompted us to find a way to connect our ideas into a system. We stumbled upon the website mathworld.wolfram.com and realized it was a valuable resource where mathematical knowledge was categorized. We found many theories and concepts that we had researched or tried to research on Wolfram. Our projects, from the inception of the firm until today, are not connected by form or style, but by the mathematical logic behind them. That's how we created the data table that WANG Fei previously mentioned. The data table, which is available on our website, represents the mathematical concepts that are potentially related to design. However, we've only included surface-level concepts that make up less than 1% of the overall data table. Our projects, which include architecture, installations, exhibitions, furniture, and more, are represented by their mathematical concepts arranged in various categories on the data table. This data sheet has become a tool that connects the dots behind our projects over the years.

I believe that math is the central source of inspiration behind our design exploration, as it represents a way of thinking and a worldview. Our ultimate goal is to continue researching and practicing the concepts on the data table for decades to come.

WANG Luming: My ultimate goal is to achieve the purity of an artist.

WANG Fei: JU Bin (Horizontal Design), your design in architecture, interior, product, yacht all explore aspects of the body, perception, and lifestyle. You have successfully integrated top resources such as lighting, atmosphere, landscape, and art. What is your ultimate goal in your work?

JU Bin: My mindset often tells me to put aside my identity and my role as an architect. From this perspective, whether we are studying design or working on practical projects, we build them upon a complete knowledge system, and through this system, we create an energy relationship with society. However, I believe this is only a phase, and from the start to the completion of a project, design is merely the outcome of knowledge, logic,

将设计的各个尺度串联、实验、对比、衍生。你们认为数学是设计的最高等级还是最终核心，抑或其他？

王鹿鸣：我的感觉大概是建筑师的工作分为两大部分，一个是作品要体现和具有时代责任，另一个是更个人化一些，就是个人喜好。社会责任感是个比较大的话题，有时也有诸多因素去束缚它的实现。个人喜好就相对自由很多，就是表达我对这个世界的认知和理解。所以数学对我们来讲可能就是兴趣、爱好，世界观的一部分——我对这个世界的理解可能很多都是基于某个逻辑的。我觉得可能也是我们作为某一个时代的人，从小就受到的基础教育慢慢把你的思路就塑造成了这样，这其实也是一种时代的产物。对于建筑师工作中相对比较见仁见智的部分，我们比较希望能够呈现出类似艺术家工作的一种状态，就是可以更多地、更自由地表达自我。

王振飞：在荷兰贝尔拉格念书的时候碰到我们的老师彼德·初摩，他给我解释他的设计课时说了一句话，他说三角形的顶点在同底边平行的线上移动的时候，形状改变，但是面积不变，我当时忽然间觉得好像左右脑连起来了，多年没有被激活的数理化的知识和一直在做的建筑设计，被一个小小的公式连起来了。因为形状改变，就可以产生很多变化，但是面积不变就能有很多的应用，我隐约感觉找到方向了。这个点可以把逻辑和设计连起来，也就是寻找数学知识背后的逻辑用来巧妙地解决建筑问题，之后就开始在这个方向上探索，毕业设计也是用这个方法研究江南水乡的院落住宅。

回国成立事务所之后有了更多的实践机会，一开始做了很多展览、装置、小家具之类，因为这类设计能够很快速地实现，让理论得到快速验证，这个还蛮重要的，因为还要有纠错的过程。这样做了几年，进行了一些数学知识向设计转化的研究和实践，但是用到的知识都比较零散，我们也没有找到把他们系统化的方法。直到2015年的时候，《城市 环境 设计》杂志邀请我们出版专辑，我们当时也已经有了一定数量的项目积累，就希望能找到一个方式把他们串联成一个系统。恰巧这时偶遇了一个网站（https://mathworld.wolfram.com），这是一个数学知识的分类网站，我们曾经研究或者尝试研究的知识都可以在这里找到。于是从开始实践到现在做的作品就可以被串起来，只不过不是以形式或风格的方式串联，而是以背后的数学逻辑分类方式串联。所以才会有王飞老师前面提到的大表格，那个表格就是我们从网站上找到的可能和设计有关的一些知识点，目前我们的工作内容只包含了一些很浅显的知识点，这些内容占整体的1%都不到。我们的每个项目都可以用项目所用到的知识点在表格上的相应位置来表达，可以是建筑、装置、展览、家具或者其他。这张表就成为串联我们这些年工作的一个线索。

我觉得数学是我们设计背后的核心线索，意味着一个思考方式和一个世界观。而终极目标就是尽量多做些研究和实践，覆盖尽量多的表格上的知识点就好，这个事需要干个几十年。

王鹿鸣：我的终极目标是可以像艺术家一样纯粹。

王飞：琚宾（水平线设计），您做的建筑、室内、产品、游艇，都是对身体、感知、生活方式的探索，也是将顶级的资源汇聚在一起，灯光、气氛、景观、艺术等等，那终极目标是什么？

琚宾：我的思维经常告诉自己放下身份，剥离自己作为建筑师的角色。从这个角度讲，我们无论是学设计还是做项目，大部分是建立在一个完整的知识体系之上，继而通过这个体系和社会产生一种能量的关系。但我认为它是一个阶段性的，一个项目从设计到完成只是那个阶

"Ju" Series, 2018 © Horizontal Design
琚系列，2018 © 水平线设计

and methods at that stage. What is more important is the process of gaining wisdom. If I can use design as a tool to understand the relationship between individuals and the rest of the world, then all additional identities become unimportant to me. So, I consider myself both a participant and an observer. As a participant, I take responsibility in managing projects, integrating resources, and interacting with society. However, I also like to act as an observer, observing and gaining insight into the world around me.

WANG Fei: WANG Zigeng (PILLS), what is the significance of narrative in design? From narrative design and exhibition to scene design for Jiang Wen's movie *Hidden Man* (2018), to space design for other exhibitions, and now as a curator, what reflections have you had on these transitions? What are your future plans?

WANG Zigeng: After going through a few transitions in life, one slowly follows a different path from others. My first transition occurred during my internship at Atelier FCJZ. Professor Yung Ho Chang asked me to work on an MIT exhibition by applying three silent films to portray 11 projects by Atelier FCJZ. At that time, there was a film called *Infernal Affairs* in Hong Kong, and Chang intended to mix the scenes of the architectural project together with this film, as if the film was taking place in this project.

Later, my graduate application portfolio was directly titled "The Narrator." At that moment, I had just graduated from college and had many thoughts that I wanted to convey, which I summarized as "encoding-decoding." The subjectivity that came out of this process was the basis for my creation.

I then took a film class at Princeton University, which was highly beneficial to me. In class, we intensively studied the symbols and shots in films, including Jiang Wen's films. I even had the chance to meet Jiang Wen at a forum at the University of Pennsylvania, and at that moment, I imagined it would be great if I could participate in Jiang Wen's films. It turned out that the year after, I worked in his crew for his film *Hidden Man* after returning to China. What a special coincidence! My first graduate studio was also film-related, and the course taught by Liam Young had a lot of influence on my creations afterwards.

So, it has been eight years since I first participated in exhibitions, and the path of our practice is quite different from that of most architecture firms. People get to know us from the exhibitions, which has brought us some opportunities as well as many issues. The opportunity is that through continuous practice, including installations, artworks, exhibition design, and curation, we will have the chance to be engaged in different media and applications in various research topics, which is a pleasure for me to work with. The challenge is that the team and the project will have less involvement in architecture, which I am still obsessed with. Thus, I am balancing these two aspects.

WANG Fei: ZHANG Shuo (SURE Design), you have been practicing architecture for more than a decade and have become the most sought-after graphic designer in Shenzhen. You have worked closely with architects on many projects. What do you think is the relationship between architecture and graphic design?

ZHANG Shuo: The phrase "sought-after" seems a bit too strong, doesn't it? We have collaborated with architects on more than 90% of our projects, working on everything from signage and environmental graphics to brand design, promotional materials, and communication graphics.

I used to be an architect and am now a graphic designer. In many of the projects we work on, I find myself switching between these two roles, sometimes acting as an architect doing graphic design, sometimes as a graphic designer, and sometimes as a hybrid of both. We have to consider the relationship between space, graphics, and the new relationship that emerges from the intersection of space and graphics. So sometimes, I feel like I am not fully in either of these two roles, but rather a third entity, like a director, directing the architect to do a part of the work first, "Architect, move

段知识、逻辑和方法综合后的结果。可我们更重要的体验是通向智慧的过程，如果通过这个载体能让我懂得个体和世界的关系，那么所有额外的身份我认为都不重要。所以我就会把自己视作一个参与者或旁观者。当我是参与者的时候，无论是工作中掌控项目、整合资源，还是和当下社会产生联系，我都会把对应的角色尽责地做好，但现在我成为旁观者的角色会越来越多。

王飞： 王子耕（镜像），叙事对设计有什么意义？从做叙事性的设计和展览，到姜文《邪不压正》电影的场景设计，再到对其他艺术的展览进行空间设计，再到现在作为策展人，这样的转变有什么思考？未来有何计划？

王子耕： 人生就是有那么几个转折，慢慢就会跟别人的路径不一样。我第一个转折可能是大五的时候，在非常建筑实习，那个时候张永和老师就让我做一个 MIT 的展览，用三部默片的形式去描述 11 个非常建筑项目，当时香港有一部电影叫《无间道》，张老师的意思是把非常建筑的项目场景跟这个电影糅在一起，就好像这个电影在非常建筑的项目里发生。

然后就是我的研究生申请作品集，名字很直接，就叫《叙事者》。那时候刚大学毕业，脑子里有很多想法想要表达，我把这些想法归纳为"编码—解码"，在这个过程里产生了主观性，这个主观性是推动创作的基础。

再到后来我在普林斯顿大学选修了电影，这个课让我受益很多，课上我记得我们一起细致地研究电影里的符号和镜头，包括姜文的电影，也在宾大的一次论坛见到了姜文，当时我想如果能参与姜文的电影就好了，结果我回国后的第二年就在他的电影《邪不压正》的剧组里工作了，我觉得命运特别巧合。我的第一个研究生课程也是和电影相关，当时选的利亚姆·杨的课程，对我之后的创作也有很多影响。

所以从我第一次参展到现在已经是第 8 年了，我们和大部分建筑事务所的实践路径确实很不同，大家对我们的认识是从展览开始的。这给我们带来了一些机遇，也给我们带来了很多问题。机遇就是通过不断地实践，包括装置、艺术作品、展陈设计和策展，你会有机会接触不同的媒介和应用，接触不同的研究议题，这对于我来讲是一种工作的乐趣。难点就是团队和项目会离建筑越来越远，我本人还是有做建筑的执念，所以也在平衡这件事。

王飞： 张烁（烁设计），做了十年的建筑设计，现在是深圳最炙手可热的平面设计师，并一直与建筑师合作，对这两者之间的关系有什么看法？

张烁： 我们跟建筑师合作项目确实非常多，跟建筑师合作的项目甚至占到了公司项目 90% 以上的比例，我们跟建筑师的合作项目从标识导视、环境图形，到建筑师公

Wu Jian'an Solo Exhibition space design, Beijing, 2021 © PILLS Architects
邬建安个展"是海，是沙丘"展览空间设计，北京，2021 © PILLS Architects

Yutian Village guide design, Shenzhen, 2018 © SURE Design
玉田村导视计划，深圳，2018 © 张烁设计

the chair up, okay, stop and don't move." Then, I say to the graphic designer, "Graphic designer, come over here, stand here, okay." "Cut!" I have to carefully coordinate these two professional roles in the project to ensure they both serve the project's maximum effect. Sometimes, one role performs more than the other, and that's where the "plot" comes in.

In summary, our relationship with architects can take many forms, such as collaboration, conspiracy, chorus, joint effort, partnership, or even a merger. Perhaps there are occasions where we are partners.

WANG Fei: HE Yong (MeetBest), the sharing economy has been a byproduct of the 2008 global financial crisis. Many architects have started designing and operating innovative activities, such as Airbnb and WeWork. What do you think is the future of behavioral design?

HE Yong: The new products produced offline, such as Airbnb and WeWork, are not only related to human behavior, but also to the economy and production modes. At this moment, the internet has had a significant impact on the behavior, production, interaction, and socialization of all people. This influence actually reduces our offline time, causing us to spend more time online and interact in novel ways. And such behavioral migration naturally affects our offline behavior. I believe that this may be the most significant social transformation in our daily lives since the introduction of the mobile internet. Our views on architecture are diverse.

In 2008, the internet sharing economy model began to influence offline behavior. With the change of this organizational form and production methods, a large number of offline spaces have also changed. When I was in the United States, I conducted some simple research and discovered some correlations of English vocabulary in the internet system, such as the words related to the type of work in the entire internet system that emerged from the original engineering field. In the field of architecture, engineers are referred to as "engineers" in Chinese, but "programmer engineers" in English. However, there is an intriguing word: "product manager." They are all-encompassing talents who comprehend the market, the design, and the technology. This may be closer to the role that some of us architects aspire to play. Through their understanding of behavior, human society, and trends, such a person can ultimately influence or respond to this era. When transitioning from online to offline, we need a large number of "product managers" who comprehend the offline systems and the behaviors of offline individuals. Through their technological expertise, they can alter the real world. People with architectural backgrounds have participated in entrepreneurial endeavors at the moment, such as Airbnb and WeWork. In fact, as architects, the way they understand or recognize many things is remarkably similar to the core role of product managers in the internet system. In the current digital age, physical spaces are continuously merging. It is meaningful and valuable for architects to respond to and even influence our physical world based on their understanding of the physical and virtual worlds.

After studying and working in western countries and regions, I have observed that in regions with more developed real estate industry or space systems, space is operated differently than in other nations. In some countries, the entire space operation is treated like a manufacturing industry where a parcel of land is regarded as a product, and its resources are appropriated. Its designer and developer then use a variety of raw materials to produce a product before selling it to another party. The entire procedure resembles the production and distribution of a product. However, a more effective space operation is one that is more like the service industry. Architects operate our urban spaces, and such design will have a timeline for continuous expansion. Designers and operators accompany it throughout the duration of the timeline. However, in many countries, architects and designers leave the building the day it is completed and have no further involvement. A qualified multi-project will get better and better as time passes. As it expands, more and more people will participate, and the overall project's quality will not decrease but rather increase. On the other hand, a large number of projects may

司的品牌设计、推广、物料和传播，都有参与。

我以前是建筑师，现在是平面设计师。在我们做的很多项目中，我的身份需要在这两个职业中不停切换，有时候是以建筑师的身份在做平面设计，有时候又单纯地只是平面设计师，有时候是混合不停地切换，可能混合的居多。我们需要同时思考空间的关系，思考平面的关系，思考空间和平面相交之后出现的新关系。因此，有时候我觉得自己在项目中这两种身份皆不是，而是另外一个身份，是第三个人，像一个导演，先指挥建筑师做一部分工作"建筑师你把椅子搬起来，好，你停别动。"再跟平面设计师说"平面设计师你走过来，你站这儿，行可以了。""卡！"我在项目中要不停地调度这两种职业身份，来确保他们都是为项目的效果最大化服务的，有时候其中一个身份表演的多一些，那也是"剧情"需要。

总之我们和建筑师的关系，有时候是合作、有时候是合谋、有时候是合唱、有时候是合力、有时候是合伙，甚至有时候是合并。

王飞 何勇（好处），共享经济是2008年经济危机时的产物，众多的建筑师从那时开启了新的行为方式的设计和运营，比如airbnb、WeWork等都是建筑师创立的。你认为对行为方式的设计的未来是什么？

何勇： 像Airbnb以及Wework这类线下产生的新事物其实都跟行为有关，而行为背后其实是跟经济以及生产模式有关系。互联网对现阶段全人类的行为方式、生产方式、互动方式以及社交方式都产生了巨大的影响。这种影响其实侵蚀了我们大量线下的时间，使我们转移到线上，用新的方式进行互动。这样的行为迁移自然也会影响我们线下的行为，我觉得这个可能是有移动互联网的概念之后对人类社会最大的改变。我们对建筑的认知是多样的。

2008年，互联网的共享经济模式开始影响人们线下的行为。大量线下的空间也随着这种组织形态以及生产方式的变化而变化。我之前在美国的时候做过一些简单的研究，发现英语词汇在互联网系统里的一些相关性，比如互联网整个系统中有关工种的词，大部分都源自于原来的工程领域。在建筑领域，engineer的中文是工程师，英语却叫程序员，建筑师architect在西方的语境是架构师，然后我们model maker在西方的语境里是创客等。但是有一个词很有意思，在他们的领域里叫产品经理，是一个懂市场、懂设计还懂技术的综合人才，这个人可能更接近于我们一些建筑师梦想成为的角色。这样的一个人能够通过其对行为的理解，对人类社会的理解，对一个趋势的理解，最终影响或回应这个时代。当线上转到线下的时候，我们需要大量的了解线下系统以及线下人的行为的所谓"产品经理"。他们通过对于技术的认知去改变线下的世界。这个时候就出现了像Airbnb、Wework这种有建筑师背景的人参与创业的项目。其实他们作为建筑师对很多事情的理解或者是认知的方式，与产品经理在互联网体系里边最核心的角色是非常像的。在我们现在所处的数字时代，物理空间正在被不断的融合。建筑师基于对物理世界以及虚拟世界的理解，反过来回应甚至是影响我们这个物理世界是很有意义和价值的。

在西方国家和地区工作和读书后，我发现在地产或者是空间系统比较发达的区域，空间运作的方式跟其他国家是不一样的。某些国家的整个空间运作方式更像一个制造业，一块土地被当作产品、资源拿过来，由设计师和开发商用各种原材料把它制造出来以后转移给了另外一方，整个过程更像一个产品的生产和转移。但是更好的空间运作方式更像是一个服务业，建筑师对于我们的城市空间是参与运营的，这样的设计会有一个持续性生长的时间轴，设计师以及运营者陪伴它经历了完整的时间轴。但在很多国家，很多设计师在建筑物出生的那天就离开了它，之后就没有任何的关系了。一个合格的复合性项目随着时间的推移会越来越好，随着它的成长，

The Collaborators of MestBest © MestBest

好处的合作方 © 好处

have reached their zenith at their birth and then go downhill. The main problem is how architects can continuously participate in space operation. Architects should not merely observe the space from a single point; rather, they should interact with the space in various ways. During this process, the subtle attempts and adjustments by architects will also produce unanticipated outcomes, which will, in turn, help architects gain a deeper knowledge of the kind and incorporate more experience and assumptions into similar projects in the future. For a good architect, each time a relatively small attempt is made to obtain feedback on their predetermined behavior, the possibility of using this behavior in the subsequent project is evaluated, and the pertinent information is then incorporated into a more finalized form. I believe that a project with social impact or behavioral feedback is more valuable than a static project.

WANG Fei: WANG Xin (Garden Architecture), unlike others, most of your effort has been focused on exploring the influence of history on the future. Can you elaborate more on this? Are you looking backward, forward, or both, or searching for a new direction?

WANG Xin: In recent years, many architecture professionals around me have asked me a few questions. One is that I don't seem to focus on the present and instead delve into history, and the other is that I don't stick to architecture and start moving into interiors, creating props and objects. What others think of me is what prompts me to reflect on myself.

What I do can be described in two ways: contemporary innovation in historical form or rebuilding the world of Chinese poetry and painting in this era. Why do I refer to the "world of poetry and painting"? Although it is related to nature, it is not just pure nature. The Chinese view of nature is a poetic and human interpretation of nature and the relationship between man and the universe. In essence, nature is a poetic narrative, which encompasses imagination, viewing, building, etiquette, and everything in daily life. In our profession, we aim to rebuild the relationship between man and nature. In fact, we are attempting to reconstruct the world of Chinese poetry and painting using the language of architecture, which is a broad and difficult topic. Although it may not exist in the physical world, it is not impossible to build a world of language or a world of form.

My work has always been very small relative to this broad topic, starting from a small corner or an object. I have set a goal for myself: before I pass away, maybe in 40 years, I want to incorporate 100 classic aesthetic situations in Chinese history into contemporary space.

The concept of "contemporary" has two aspects. First, it is to make it architectural. For example, our studio has a project called "Blue Wave Treading." This refers to the ancient ideal of traditional Chinese people, where people can tread on waves with magic power. This seems unrelated to architecture, but from the standpoint of architecture, it is necessary to solidify this imagination and moment through the language of architecture and capture the essence of poetry. Second, contemporizing is to combine the classical aesthetic context with commercial behaviors in real life and to realize the possibility of integrating the traditional contextual model into daily life or transforming it into commercial value.

WANG Fei mentioned a second question earlier, asking whether I look forward or backward. Let me explain. For example, gardens flourished only after urbanization and exist as an aspiration and compensation for the city. If we look at tradition and the future in the context of this complementary relationship between gardens and cities, tradition and the future must also exist in a compensatory way with each other. The further we step into the future, the more we may crave tradition, like the coexistence of initiation and detachment, the back and forth, and the coexistence and integration of Confucianism, Buddhism, and Taoism. Therefore, in my heart, the past, present, and future are always mixed and intertwined. Tradition is also a component of the future, a repository of tradition for the future, a healing touch, and a brake for the future. Nowadays, in a society that is running wildly forward, we cannot just have a throttle, but we also need a brake.

越来越多的人会参与进来，整个项目的质量是不降反增的。但是在很多地方大量的项目在出生的那一刻可能就达到顶峰了，之后一定是走下坡路的。这当中的问题是建筑师如何持续性地参与到空间的运作里面。这样的参与并不是简单的用单一的角度去观察，建筑师会跟空间产生不同方式的互动。而这个过程中，建筑师细微的尝试与调整也会对项目产生意想不到的结果。这种结果会帮助建筑师对这一类事物产生更深入的理解，在下一次做同样类型的项目时就会代入更多的经验以及预设了。好的建筑师每一次用一个相对比较小的尝试去获取他预设行为的一个反馈，再通过观察这个行为的可能性运用在下一个项目里面，然后把相关的信息整合到一个比较终极的结果形态。这个时候它所产生的社会影响，或者是行为的反馈，我觉得可能会比一个非常静态的项目有更大的价值。

王飞：王欣（造园建筑），和其他在座的设计师不同，您可能大部分的精力都是在从历史中去探寻对未来的影响，能展开谈谈吗？您是在向后看还是向前看，抑或两者都是，或者是新的方式？

王欣：建筑圈的很多朋友近年来对我有几个疑问：一个是我好像并不直面当下，跑到历史中去了。二是我并不直面建筑，开始转向室内了，还做了道具器物等。别人怎么看我，正好是一面镜子，也促使我自我审视一下。

我所做的工作往小处说是历史形式的当代创新，往大处说是在这个时代重建中国的诗画世界。为什么指向

Bowl Mountain © Garden Architecture
碗山 © 造园建筑

"诗画世界"？可能说法也不太一样，也许以前我们常说要重回自然，实际上中国人所说的自然虽然与自然界有联系，但其实根本不是纯自然。中国人所说的自然，是诗化的人对自然的看法，以及人与天地万物的关系。所以，简单的讲，自然就是诗画叙事。这是一个最大的事了，涵盖了想象、观看、营造、礼仪、日常的一切。我们这个专业，重建人跟自然的关系，实际上是要试图用建筑的语言来重建中国人的诗画世界，这事很大也很难，也许不是一个真的存在的物理世界，但建立一个语言世界或是一个形式世界并非不可企及。

相对于这个有点大的事情，我做的工作一直都特别小。从一个小小的角落开始，或者从一件器物开始。我大概给自己定了一个目标：在我倒下之前，可能剩下40年吧，把中国历史上的100个经典美学情境都当代化、空间化。

这个"当代化"，一方面是把它建筑化，举个例子，我们工作室有一条线索叫"碧波踏浪"。碧波踏浪是传统中国人自古的理想，指的是人可以得仙术踏浪而行，似乎它跟建筑是没有关系的。但在建筑的立场当中，需要把这种想象和瞬间通过建筑的语言凝固下来，定格诗意。另一方面，就是把经典的美学情境跟现实生活和商业行为结合在一起，实现把传统的情境模式融入日常，或转化为商业价值。

刚才王飞老师提到第二个问题，是向前看还是向后看的问题。我解释一下，比如说园林，是因为城市化开始之后园林才有真正的兴盛，园林是作为城市的一种渴望和补偿性的存在。如果拿园林和城市这种互补关系来看传统和未来，一定也是相互补偿性的存在。我们越走向未来，可能越渴求传统，就好像入世与出家是并存的，是来回的，比如说儒释道的并存与融合，是三位互补性共生。所以在我心里，不存在是向后看还是向前看的问题，过去、现在、未来，三者永远是交混交织在一起的，传统也是未来的组成部分，传统是未来的思想库，也是疗愈的怀抱，传统是未来的刹车，当下这个在往前狂奔的社会不能只有油门，也一定要有刹车。

王飞：如果让你对现今的建筑教育做些改变，有何建议？

马圆融：我觉得可以在教学中增加更多的真正和真实材

WANG Fei: What suggestions do you have for changing the current architectural education?

MA Yuanrong: I think we can incorporate more training with real materials, conduct more materialistic research, and integrate more hands-on design methodology into teaching. Nowadays, many courses are digital which could largely weaken our ability to experience and perceive reality.

JU Bin: Practice more. Architecture is a profession that relies on practice to testify and verify.

WANG Zhenfei: I recommend adding more discussions on logical thinking in the current pedagogy, not only in architecture but also in other majors. It is important to help students think more about the relationship between geometry, function, and forms, which is more direct and easy to understand. It would help students establish a clear logic behind their thinking, which is an essential asset for an architect. It should also help students who think more mathematically.

WANG Zigeng: The attractiveness of architecture is that it satisfies a certain curiosity or ability to construct and organize the material world. The foundation of this discipline is built on physical construction, so for a long time there must be a disciplinary core to support it, such as construction, materials, and geometry, which are all core issues. However, is it enough to solve only the core problems to complete a building? Obviously not. Architecture is also a complex game of social organization, communication, marketing, cost, and different interests. A good architect must also be the coordinator of social resources. In addition, the material basis of architecture is constantly changing. Over the centuries, under distinct cultural contexts, although we say that we are building houses, the aspirations for technology, demand, organization, core elements, and spatial are changing. Architecture must answer a question of cultural significance in the time dimension: Why build? How to build? Therefore, architecture is linguistic and intermediate at the same time, not only in the sense of construction. This is the problem that cultural criticism and technical history must solve. Domestic architectural education is too homogeneous and neglects many related dimensions, thus ignoring the possibility of connecting architecture to the outside and expanding its own capabilities.

WANG Xin: In short, it means restarting the education of interest. On the one hand, it revolves around people's lives and emotions, arousing interest, making things come alive and driving people to pursue a career in architecture. The second aspect is to enhance the cultivation of imagination and generate more possibilities for cross-disciplinary work. The third aspect is to strengthen the education of reality and in situ. "Reality" means education should not be separated from social reality, and "in situ" means education should not be detached from people and life circumstances.

ZHANG Shuo: Maybe it's because I basically studied on my own during my college years without any special professional training or education, so I don't know what a professional architectural education should be like. But like graphic design, I feel that the basic education of architecture and graphic design in college is very vague, and after I graduate and put myself to work, I basically need to relearn it all over again. I suggest that students need to enter professional practice early in their professional education.

料相关的一些训练，更多物质性的材料研究，融入更多动手的设计方法论。因为现在很多的课程基本上都是数字化的，我觉得这个会很大程度上削弱我们体验和感知现实的能力。

琚宾： 多实践，建筑是通过实践去验证和证实的专业。

王振飞： 建议从本科入学就开始在教学的过程中多加入一些关于逻辑的讨论吧，建筑类的，非建筑类的都好。多引导学生研究数学，几何与功能、形体的关系，这个比较直接也比较容易理解，能够帮助学生建立逻辑清晰的思维模式，这对一个建筑师来讲还是挺重要的，应该能够帮助到着重理科思维的学生。

王子耕： 我觉得建筑学吸引人的地方是他满足了建构物质世界的某种好奇心或者知识组织能力。这门学科的基础是建立在物质建造上的，所以长久以来，一定是一门学科核心去支撑他，建造、材料、几何，这些都是核心问题。然而把房子真正实现出来，是不是只解决核心问题就可以呢？显然不是的。建筑学同时也是一种复杂的社会组织、沟通、营销、成本以及不同利益方之间的博弈。一个好的建筑师，也必须是社会资源的协调人。另外，建筑学的物质基础其实也在不断的变化，几百年来，不同文化背景下，我们虽然笼统地说都是在盖房子，但其实技术、需求、组织方式、核心元素、空间诉求都在变化，建筑学还要回答一个时间维度上文化意义的问题，为什么建造？如何建造？所以建筑学同时也是语言上的、媒介上的，而非建造意义上的。这里就是文化批评、技术历史要解决的问题。国内的建筑教育指向过于单一化，忽视了很多维度，从而也就忽视了建筑学连接外部、扩展自身能力的可能性。

王欣： 简单的说，就是要重启情趣的教育，第一方面是围绕人的生活和情感，引发兴趣，因为兴趣，事情才能变活，一件苦累的事业才能终身投入。第二方面就是加大想象力的培育，产生更多的学科交叉的可能性。第三方面就是要加强对现实和现场的教育，"现实"即教育不要脱离社会现实问题，"现场"即教育不要脱离人和脱离生活情境。

张烁： 可能是因为我大学期间的学习基本是靠自学，没有受到特别专业的训练和教育，因此我也不知道专业的建筑学教育应该是什么样的。但是和平面设计一样，我感觉大学里建筑学和平面设计的基础教育都深度不足，毕业投入工作之后，基本需要靠自己重学一遍。我建议在专业教育中还是需要让学生早点进入专业实践。

Horizontal Studio
Windrider 1
Shenzhen, China 2017-2020

水平线设计
御风者 1 号
中国，深圳　2017-2020

pp.30-33: Exterior view of Windrider 1, © Xufeng Jing.

第 30-33 页：御风者 1 号室外照片，© 井旭峰。

Windrider 1 has come to life in just four years, making the impossible possible. With the best technological support and resources from around the world, it breaks the boundaries of spatial design by bringing together the best elements of where we stand on land and where the ocean flows. A great yacht design speaks of more than just luxury or extravagance. Instead, it exudes a sense of solemnity and calmness through its purity, honesty, and fineness. The phoenix-eye-shaped opening on the prow, the gill-like notches on the sides, and the champagne finish all lend grace and lightness to this 44.7-meter-long catamaran yacht, making it a piece of moving architecture on the ocean.

The idea of this yacht design was born from our client's desire to create a super yacht designed and built in China that reflects the unique lifestyle and values of China while incorporating every cultural element of this part of the Eastern world. Eventually, there are spaces solely dedicated to tea and board games, merging emotions and cultures with functionality. The yacht itself becomes a carrier that combines humanity, culture, and the ocean.

The yacht's four floors offer a combination of activity and rest areas, including tea-game living

历时四年的御风者 1 号成就了许多不可能变为可能，拥有了中国乃至全球最好的技术支持、设备资源，克服了陆地空间设计到海上空间设计的施工差异，好在设计的情感是相通的，游艇可以不是传统的奢华，她可以是贵气且纯粹、安静、内敛，质朴中有着一种出挑的气韵。船头凤眼般的开口，两侧鱼鳃板的切口，独有的香槟色的外观，都让总长 44.7m 的双体游艇力量里透出轻盈，她就是一个活动的"建筑"置于蓝色自然中。

起初开始设计游艇的机缘也来源于业主想要打造一艘中国设计、中国建造，独属于东方人生活方式的超级游艇，于是游艇里出现了茶空间、棋牌空间，在地性的功能置入像一种情感的粘合剂，连通人与人、人与海洋、人与人文，链接艺术、思想和情感。

This page, above: Outdoor deck of third floor. This page, below: Outdoor deck of first floor. Opposite: Top view of Windrider 1, © Xufeng Jing.

本页，上：单层甲板空间。本页，下：一层甲板空间。对页：御风者 1 号顶视图，© 井旭峰。

1. Outdoor Leisure Area
2. Outdoor Shower
3. Barbecue Area
4. Leisure Area
5. Water Bar Area
6. Bar
7. Outdoor Lounge
8. Jacuzzi
9. Outdoor Leisure Area
10. VIP Leisure Area
11. Food Preparation
12. Suite
13. Public Restroom
14. Public Corridor
15. Suite Bathroom
16. Captain's Bathroom
17. Captain's Room
18. Cockpit
19. Outdoor Platform
20. Living Room
21. Kitchen
22. Game Room
23. Tea Room

1. 户外休闲区
2. 户外淋浴
3. 烧烤区
4. 休闲区
5. 水吧区
6. 吧台区
7. 室外躺榻
8. 泡池
9. 户外休闲区
10. VIP 休闲区
11. 备餐区
12. 套房
13. 公共卫生间
14. 公共过廊
15. 套房卫生间
16. 船长卫生间
17. 船长室
18. 驾驶舱
19. 户外平台
20. 会客厅
21. 厨房
22. 棋牌室
23. 茶室
24. SPA

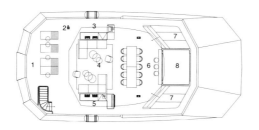

Third floor plan / 三层平面图

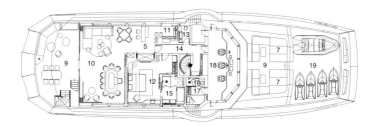

Second floor plan / 二层平面图

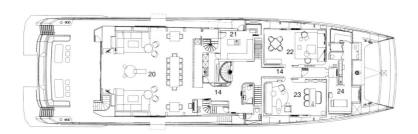

Basement floor plan / 负一层平面图

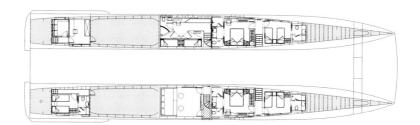

Basement floor plan / 负一层平面图

Opposite: Outdoor deck outside control room, © *Xufeng Jing.*

对页：驾驶舱室外甲板，© 井旭峰。

This page, Opposite: Exterior view of Windrider 1, © Xufeng Jing.
本页，对页：御风者 1 号室外照片，© 井旭峰。

Section A / 剖面图 A

Section B / 剖面图 B

rooms, bedrooms, suites, and entertainment areas. The color and material choices, such as shades of grey, leather, and beige linen, as well as light champagne-colored painted panels, reflect the modesty and Zen of Eastern culture, giving the yacht an honest and frank character. Additionally, all these materials can be assembled individually in pieces and butt joints.

In addition to literary and experiential perspectives, the design also incorporates the placement of art pieces that reflect the surrounding nature. It showcases a performance of lifestyle, values, and thinking. These art pieces range from drawings by ZHAO Wuji to works by JIN Yuyi, which reflect the endless dreams of deaf and mute children towards the infinite ocean. Contemporary art presents different themes in various stage settings, emphasizing the uniqueness of the space it occupies.

Windrider 1 blends the charm of both Chinese and Western cultures, making it a stunning salon above the ocean.

This page, above: Tea house. This page, below: Master bedroom of second floor. Opposite, above: Reception area of first floor. Opposite, below: Main circulation of first floor, © Xufeng Jing.

本页，上：茶室。本页，下：二层主卧室。对页，上：一层交通节点。对页，下：一层接待空间，© 井旭峰。

Credits and Data
Project title: Windrider 1
Client: Private
Location: Shenzhen, China
Completion: December 2020
Principle Partners: BIN Ju, PAN Qinchao / Horizontal Design
Design Team: Guo Dayu, HU Yao, LEI Xingyue
Technical Support Team: NIE Hongming, HU Kai, MO Zhibing, LI Junhua, ZHANG Luokai, HUA Ruixiang
On-site Designer: HU Kai
Soft Decoration/Material Team: DENG Yangwen, QIN Jiongbin, LUO Qiong, CHEN Yuhua, LU Yuefei
Project Owner: China Cup International Regatta
Project Manufacturer: Heysea Yachts Company Limited
Project area: 1700 m²

　　船体上下分成动静分离的茶·棋会客区、卧室区、套房区、飞桥娱乐区四层，材料的选择更贴近东方的宁静调性，也更纯粹耐读，深色浅色的灰影、米白色的皮料麻料、浅香槟色的漆面板，都进行有机对缝的独立分块装配，都可独立拆卸。

　　空间内除了体验性、文学性的解读，还有艺术品的摆放，这些都是一个与自然相互呼应并彼此加分的大氛围的营造手法，是一种生活方式甚至思考方式的引导。墙上挂的艺术品，从赵无极的绘画到金羽翼的作品，反映聋哑儿童对大海的无限遐想。当代艺术在不同空间以不同主题模式呈现，矛盾和多元的气质在这里共存共生，指向了空间里的唯一性内核。

　　御风者1号，她是具有国际化性质同时兼具中国东方文化气韵的游艇，蓝色海域上的会客厅。

FUN Design
Grand Gourmet Flagship Store
Shanghai, China 2017

凡谙设计
Grand Gourmet 旗舰店
中国，上海 2017

For centuries, Foie Gras, a renowned traditional French delicacy, has been considered one of the three most exquisite treats in the world, along with truffles and caviar.

The Grand Gourmet flagship store is situated in Shanghai Xintiandi and boasts a parametric Art Deco style inspired by the colonial culture of Shanghai. It functions not only as a shop but also as a gallery, communication space, and event space, showcasing products, offering a lifestyle experience, and facilitating social interaction.

The main display area is located in the east part of the shop, with various display strategies employed to attract customers' attention. The central table serves as a display table for Foie Gras and can transform into a dining table during tastings, a buffet table for exhibition openings, or a display area for art pieces. A foretaste area is situated in the western part of the shop, which can be converted into a bar during events. The shop space remains flexible and adaptable to upcoming events.

The store has combined modern technology with traditional handcrafts. The fractal hexagon pattern lends the store a distinctive identity and organizes all functions. The construction process involved cutting-edge technologies, such as 3D printing, laser cutting, and 3D milling, while respecting traditional hand-made processes for Foie Gras, demonstrated through the hand-embedded brass bar on the floor, hand-polished brass panels, and hand-casting display tables.

The creative fusion of traditional handcrafts and modern technology gives the store a unique character, reminiscent of how Foie Gras blends old hand-making processes with modern lifestyles.

pp.42-43: Shop Interior.
This page: Shop Exterior. Opposite: Shop Interior.
第 42-43 页：店铺内景。
本页：店铺外景。对页：店铺内景。

1. Display Area
2. Tasting Area
3. Kitchen
4. Office

1. 展示区
2. 品鉴区
3. 厨房
4. 办公室

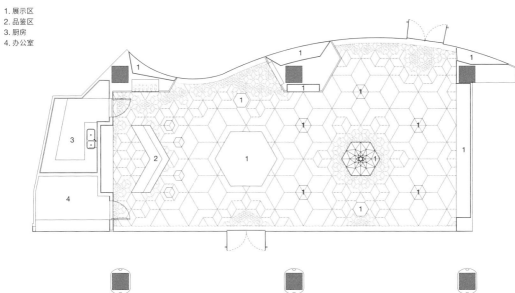

Floor plan / 平面图

This page, above, below left: Door Handle. This page, below right: Display Cage. Opposite: Shop Interior.

本页,上,左下:门把手。本页,右下:展示笼。对页:店铺内景。

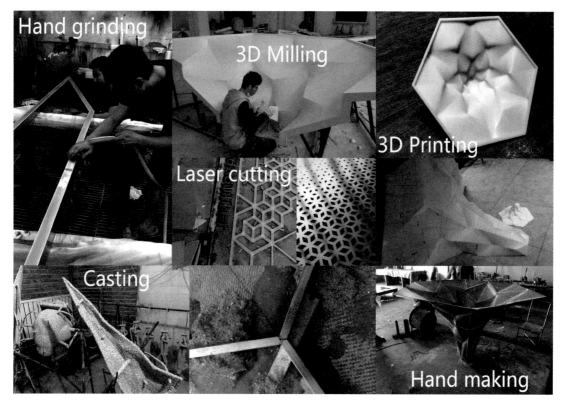

This page: Construction Process. Opposite, above: Display Stand. Opposite, below: Tasting Area.
本页：建造过程。对页，上：展示台。对页，下：品鉴区。

几个世纪以来，著名的法国传统美食鹅肝酱被公认为是世界上最精致的三种美食中的佼佼者，另外两种是松露和鱼子酱。

Grand Gourmet 旗舰店位于上海新天地，参数化 Art Deco 的风格源于上海的海派文化。作为对公众展示的重要窗口，它的功能不仅是一间店铺，还是艺术馆、交流空间和活动空间。人们在这里品尝和购买鹅肝酱，参观艺术展，参加 Grand Gourmet 的各类活动，互相交流，体验不同的生活方式。

Credits and Data
Project title: Grand Gourmet Flagship Store
Year of Design: 2016
Year of Complete: 2016
Location: Xintiandi, Shanghai, China
Floor Area: 161 sqm
Architect: FUN
Photographer: WANG Zhenfei

商店空间最大限度地保持灵活，以适应不同活动的需要。店内陈列功能主要位于东侧，采用各种陈列策略来吸引顾客的注意力。中间的大桌子平时作为鹅肝酱的展示台，也可以在品鉴会时变成餐桌，展览开幕时变成自助餐台，展览期间变成艺术品展示区。店铺西侧设计了一个品鉴区，活动期间将变成酒吧。

店铺结合了前沿的设计、制造技术和传统的手工艺，分形六边形图案赋予商店强烈的标识性，同时组织了所有功能。施工过程涉及许多新技术，例如 3D 打印、激光切割和 3D 雕刻等。为了表达对纯手工鹅肝酱的尊重，很多传统手工艺也被应用，比如纯手工镶嵌铜丝水磨石地面、纯手工打磨的铜板、纯手工翻模铸造的展台等。

创造性地将传统手工艺与先进的设计和建造技术相结合，使得店面与众不同，正如鹅肝将古老的手工制作工艺与现代生活方式相结合。

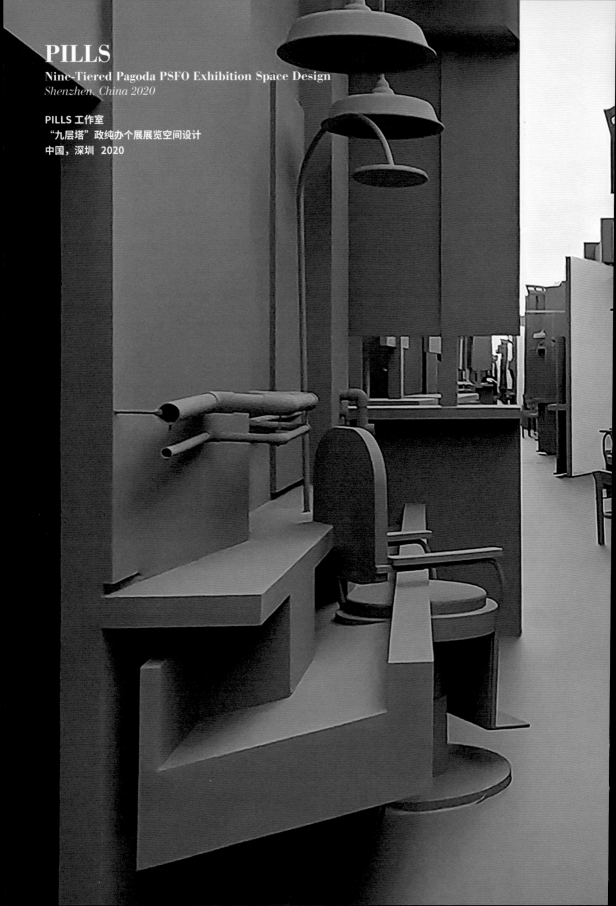

PILLS
Nine-Tiered Pagoda PSFO Exhibition Space Design
Shenzhen, China 2020

PILLS 工作室
"九层塔"政纯办个展览空间设计
中国，深圳 2020

Nine-Tiered Pagoda: Spatial and Visual Magic comprises of nine exhibitions. As the first project in the series, it is a spatial presentation based on the works of the art group PSFO. In this exhibition, PSFO presents a digital composite portrait of Mr. Zheng, which personifies the idea of "We become I" from the five members of the group. The exhibition responds to Mr. Zheng with a collage of five typical collectivist spaces - water house, canteen, screening room, barber shop, and bathhouse—creating a blue corridor to showcase PSFO's artworks of diverse mediums. The exhibition extracts and juxtaposes different collectivist spatial elements to form a new "species" of coincidences and relations. The blue corridor includes parallel mirrors at both ends, generating the visual illusion of infinite extension and accommodating detached contents. The installation creates a dream-like atmosphere that transports the viewer to a past that seems familiar yet unexperienced.

pp. 50-51: The blue tube-shaped corridor, © PILLS Architects. This page, Opposite, above: The blue tube-shaped corridor. This page, below: Looking at the corridor from the gallery, the size of window is similar to the paintings of PSFO, © PILLS Architects.

第 50-51 页：展览现场蓝色筒子楼廊道，© *PILLS* Architects。
本页，对页，上：展览现场蓝色筒子楼廊道。本页，下：从画廊展厅望向走廊，画廊区域的窗洞切割与政纯办画作尺幅相似，© *PILLS* Architects。

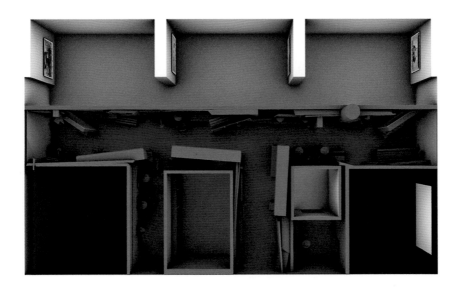

Top view of 3D mode / 三维模型顶视图

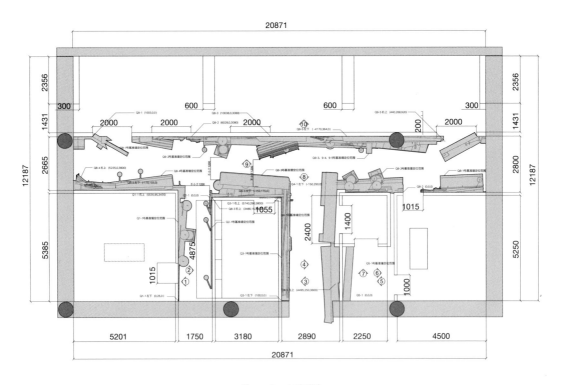

Floor plan / 平面图

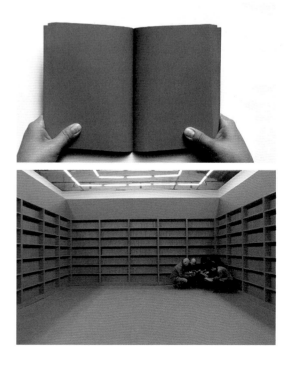

This page, above: Exhibition room of Mr. Zheng. This page, below: Library, PSFO, © PILLS Architects.

本页，上：政纯办作品"政先生"画像展厅。本页，下：政纯办作品"图书馆"，© PILLS Architects。

"九层塔：空间与视觉的魔术"由9个不同形式的展览组成。作为"九层塔"的第一个项目，本次展览以艺术小组"政纯办"的作品为基础进行空间呈现。作品空间概念来源于政纯办的数码合成肖像作品"政先生"，"政先生"的头像是五位艺术家面部特征的集合——以"集体"的五官拼贴出一个不存在的"个人"。相对应地，我们把五种集体主义典型空间（水房、食堂、放映厅、理发馆、澡堂）和关于"物"的符号嵌套组合，去促发偶然性，形成杂交的物种，从集体中生发出新的个体。在蓝色的筒子楼廊道空间，串联政纯办不同媒介作品的展示需求。廊道两端的平行镜面使视觉在有限的空间内产生无限延伸的错觉，创造出抽离内容的形式游戏场所，将观者带入似曾经历过的从前，也带入未曾经历过的梦境。

Opposite, above: Looking from the exhibition space to the blue tube-shaped corridor. Opposite, below: The entrance to the exhibition space. This page, above: At the opening of the exhibition, the artist leaning on the right side of the wall is Hong Hao, a member of the Polit-Sheer-Form Office. This page, below: Looking from the Library room to the exhibition space, the window openings reveal benches cut through the borders and paintings covered by blue baffles, © PILLS Architects.

对页，上：从画廊区域望向蓝色筒子楼廊道。 对页，下：展厅入口。
本页，上：展览开幕式现场，右侧倚墙者为政纯办成员艺术家洪浩。
本页，下：从"图书馆"展厅望向画廊展厅，透过窗洞可以看到被窗口边界切开的长椅和被蓝色挡板遮盖的画作，© PILLS Architects。

Credits and Data
Project Title: PSFO Solo Exhibition Unity is Strength Spatial Design
Exhibition: Nine-Tiered Pagoda: Spatial and Visual Magic
Curator: CUI Cancan
Producer: LIU Xiaodu
Event: 24 October—29 November 2020
Location: Pingshan Art Musuem, Third Floor
Exhibition Space Design: PILLS Architects
Exhibition Space Architect: WANG Zigeng
Exhibition Space Design Team: YAN Yu, SHI Jianli, PAN Jingxian, ZHOU Fangda, XU Shiman, DU Ximing, ZENG Haochun
Completion Year: 2020

PILLS
Nineteen Ninety-Four
Beijing+Shenzhen, China 2021

PILLS 工作室
1994 年
中国，北京 + 深圳 2021

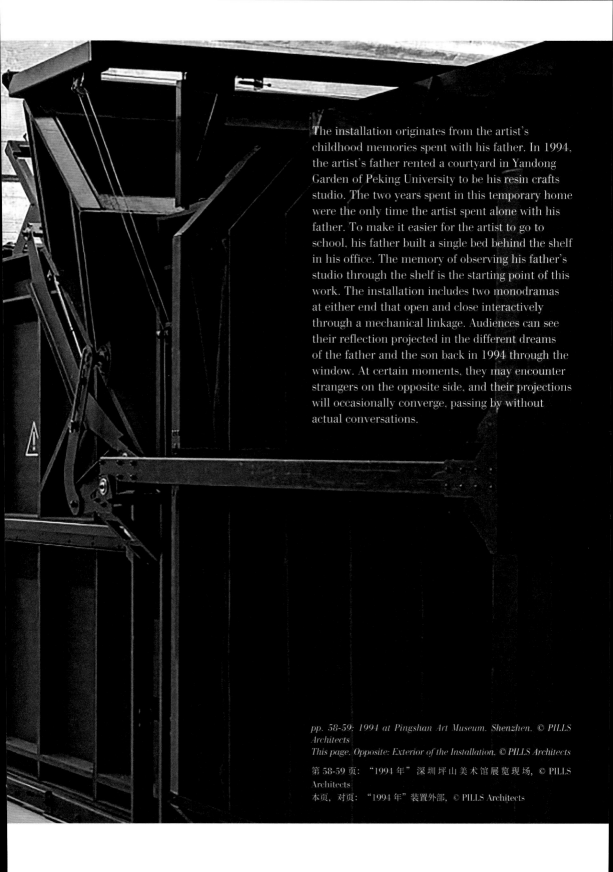

The installation originates from the artist's childhood memories spent with his father. In 1994, the artist's father rented a courtyard in Yandong Garden of Peking University to be his resin crafts studio. The two years spent in this temporary home were the only time the artist spent alone with his father. To make it easier for the artist to go to school, his father built a single bed behind the shelf in his office. The memory of observing his father's studio through the shelf is the starting point of this work. The installation includes two monodramas at either end that open and close interactively through a mechanical linkage. Audiences can see their reflection projected in the different dreams of the father and the son back in 1994 through the window. At certain moments, they may encounter strangers on the opposite side, and their projections will occasionally converge, passing by without actual conversations.

pp. 58-59: *1994 at Pingshan Art Museum, Shenzhen.* © PILLS Architects
This page, Opposite: *Exterior of the Installation.* © PILLS Architects
第 58-59 页:"1994 年"深圳坪山美术馆展览现场,© PILLS Architects
本页,对页:"1994 年"装置外部,© PILLS Architects

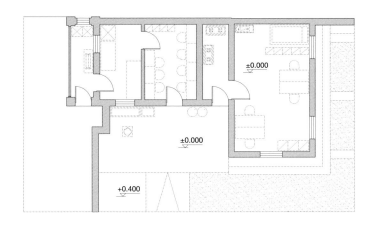

Floor plan of father's studio in 1994 / 1994 年父亲工作室的平面图

This page: Smoky mirror world. Opposite, above: Reception area of first floor. Opposite: Space of the father and son, © PILLS Architects

本页：烟雾缭绕的镜中世界。对页：父与子的空间，© PILLS Architects

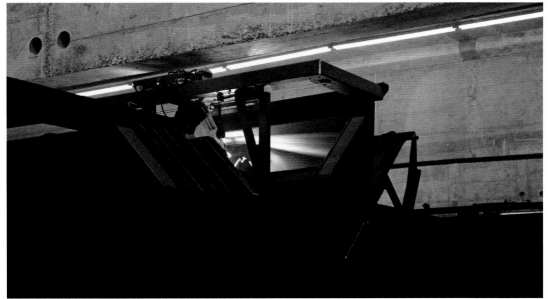

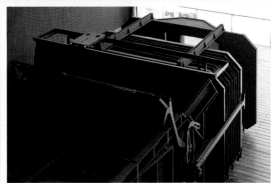

Exterior detail of the installation, © *PILLS Architects*

"1994年"装置外观局部，© PILLS Architects

这个装置缘起于作者小时候和父亲共同度过的一段时光。1994年，作者父亲租下北大燕东园的一个院子当作办公室，制作树脂工艺品。在这个临时居所的两年是作者仅有的和父亲单独相处的时光。为了方便上学，作者的父亲在他办公室的货架后搭了一个单人床，透过货架观察父亲办公室的记忆成为这个作品的起始点。两个端头的单人剧场通过联动机械交互开合，透过窗口，观众可以看到自己的虚像置身于1994年的父与子不同的梦境里，也可以在某一个刹那与对面的陌生人不期而遇。彼此的影像会偶然性地出现在同一个空间内，一晃而过却无法交流。作品通过机械、影像、交互、置景等手段的共同作用，将记忆的片段再现，并通过光学反射原理将不同的虚像并置在场景空间中，形成与现实空间关联却又独立的叙事空间，并在其中创造偶然性带来的叙事层次。这件作品起始于作者个人的父子回忆，同时也希望探讨亲缘关系中的依赖、对立、和解和遗憾的轮回。

Credits and Data
Project Title: Nineteen Ninety-Four
Location: Beijing, Shenzhen
Commissioner: Wind H Art Center, Pingshan Art Museum
Completion: 2021
Principle Designer: Zigeng WANG
Design Group: YU Jun, HAN Jianye, WANG Manying, YAN Yu, ZHOU Fangda, YU Yang, ZHANG Tiezhi, MA Cheng, ZHU Huiran, LIN Yao

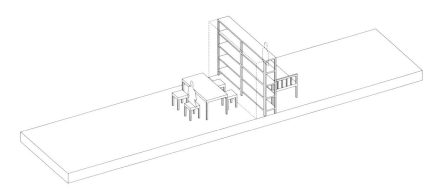

Axonometric drawing of the physical installation / 装置轴测图

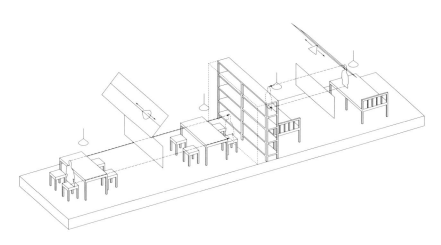

Axonometric drawing of installation with hidden spcaes / 装置隐秘空间的轴测图

1. Mirror
2. Camera
3. Audience A
4. Mirror Projection Light Source
5. Mirror Projection Screen
6. Primary Light Source
7. Mirage Imaging Screen
8. Audience B's Virtual Image
9. Audience A's Mirror Projection
10. Supplemental Light Source
11. Audience A's Virtual Image
12. Audience B's Mirror Projection
13. Audience B

1. 镜子
2. 摄像头
3. 观众 A
4. 镜像投射光源
5. 镜像投射界面
6. 主光源
7. 幻影成像界面
8. 观众 B 虚像
9. 观众 A 镜像
10. 补光光源
11. 观众 A 虚像
12. 观众 B 镜像
13. 观众 B

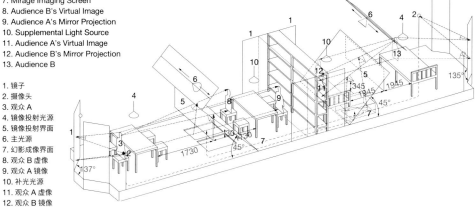

Axonometric drawing of installation with hidden spcaes and relfections / 装置隐秘与反射空间的轴测图

Garden Architecture
Little Cave-Sky
Hangzhou, China 2021

造园建筑
小洞天
中国，杭州 2021

叠透洞遠

Dongtian, Cave-Sky (a type of Chinese landscape embodying unique worldviews and values that emphasize the harmony between humanity and nature. Dong: Cave; Tian: Sky) is one of the main focuses in our teaching experiments. With the exploration in teaching, we have opened up new possibilities in practice.

Xiaodongtian, Little Cave-Sky (a studio built for collectors obsessed with viewing rocks, named after an inhabitable rock) is a contemporary experiment of the Daoist concept of "Dongtian" in China, creating a unique time and space in a small 60-square-meter apartment, a tiny internal world. In today's cities, Little Cave-Sky is particularly necessary, nesting like a wormhole in a mysterious dream, transcending the outside world, telling its own story and value. The cave is a way of hiding oneself in the city.

Little Cave-Sky has no external form; it is an introspective way of viewing mountains and water. Therefore, its form is surrounding, emergent, and embodied. No rooms or fixed functions exist, and daily activities are discovered according to the surroundings. There are no furnishings, only the

hint of the elevation difference to the human body, which has nothing to do with comfort but awakens the poetic memories of mountains and waters in the body. There is no indoor nor outdoor because weather, sun and moon, and history converge here in a pictorial way. Its structure is derived from the grammar and imagination of Chinese poetry.

Little Cave-Sky experiment has achieved a combination of caves and buildings, which is not only the ideal construction pursued by ancient people but also the problem that contemporary Chinese local architecture must face: how to integrate nature with buildings? What did architecture gain from nature?

pp.66-67: Layered depth of caves, viewing the entrance from the tea house.
This page, Opposite: Viewing multiple directions: the coming path (middle), up to the sky (left), hidden depth (right).

第66-67页：叠透洞远，自茶室望向入口。
本页，对页：指多个方向，来路（中）、上天（左）、深藏（右）。

洞天，我们一直将此作为教学实验中的主要线索之一，这不仅是中国山水中的一种重要类型，更是一种独特的世界观与价值观。有了教学的探索，便开启了实践的可能。

小洞天，是为痴迷于赏石的藏家所建的工作室，这间工作室就是一块巨大的山石，让主人住进他所爱的石头中去，遁入他的时间里，遁入他的世界中。这块可以居住的山石，我名之曰"小洞天"。

小洞天，是对中国道家"洞天"思想的一次当代实验，在仅有 60 平方米的公寓中营造一个别样的时空，一个极小的内部世界。在如今的都市中，小洞天愈加必要，如虫洞一般嵌套在某处，如一场神秘梦境存在一时，超然于世外，闭门即是深山，独立叙事，独立价值。洞，是一种隐居于城市中的方式。

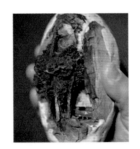

小洞天，没有外形，是一种山水的内观方式。因此，他的形态是周遭的，涌现式的，是身体化的。没有房间，没有固定的功能，起居坐卧皆在于发现。没有家具，只有高差之于人体的暗示，这无关舒适度，这是一种对山水记忆之于肢体上的诗画般的唤醒。没有室内外，因为天候日月历史皆以画意的方式汇聚于此，他的构造源自于中国诗歌的语法和想象。

小洞天的实验，完成了一次洞与房的合体，这不仅是古人所追求的理想营造。更是当代中国本土建筑学要面对的问题：自然如何与建筑合体？建筑究竟从自然中获取了什么？

Credits and Data
Project Title: Little Cave-Sky
Location: Hangzhou, China
Completion: 2021
Principle Designer: WANG Xin, SUN Yu

This page, above: Viewing the bedroom from study room. This page, middle: Viewing study room from the bedroom. This page, below: The crack between the tea house and the atrium of the cave. Opposite, above: The overview at the atrium. Opposite, middle left: "Worm's hole" of Taihu Stone. Opposite, middle right: "The peach blossom spring" one the tea pot. Opposite, bottom left: Cave sky in Juqu (Old Taihu) Forest Scroll by WANG Meng, Yuan Dynasty. Opposite, bottom right: Cave sky by XIAO Yuncong, Ming Dynasty.

本页，上：自书房望向卧室。本页，中：自卧室望向书房。本页，下：自茶室通向洞之中庭的"裂缝"。对页，上：洞的中庭全景。对页，中左：太湖石袖峰的"虫洞"。对页，中右：壶壁上的"桃花源"；对页，左下：王蒙的洞天，《具区林图卷》，元代。对页，右下：萧云丛的洞天，明代。

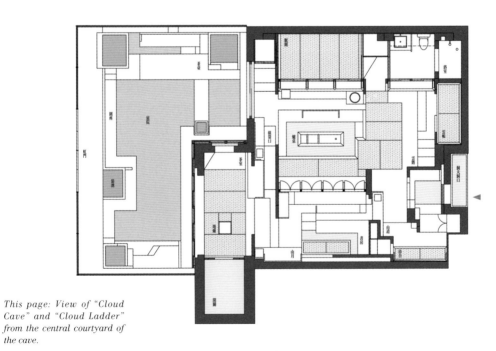

This page: View of "Cloud Cave" and "Cloud Ladder" from the central courtyard of the cave.

本页：自洞中庭望去"云窟"与"梯云"。

Floor plan / 平面图

Elevations / 立面图

Garden Architecture
Asia Bamboo Life and Art Exhibition
Hangzhou, China 2019

造园建筑
"东方竹"亚洲竹生活艺术展序幕空间设计
中国，杭州 2018

This is the prologue space designed for the Asia Bamboo Life and Art Exhibition. The bamboo forest and bamboo waves are used like the beginning of the dreamlike experience, allowing the audience to immerse themselves directly in the theme through a situational approach. The Eastern nature of bamboo is directly materialized into a garden scene that can be entered.

The bamboo forest is a natural sanctuary in China and a gathering place for literati. Bamboo waves represent a form of formalized and landscaped music from the XIAO flute, creating a cohesive and harmonious atmosphere. Bamboo waves also serve as a natural exhibition platform, providing a unique background for exhibits and contributing to the display's commentary and ambiance.

It is an experiment in bringing contemporary Chinese gardens into museum exhibitions, enlarging the experience of the exhibition and the visual field of perception. At the same time, this creation is also a building experience course of the School of Architecture at China Academy of Art, which is open to the public and serves as an artistic and educational activity centered around the theme of nature and created naturally.

pp.74-75: Opening of the exhibition—the embraces by nature.
This page: The first chapter of the exhibition, the bamboo waves surge towards the bamboo forest. Opposite: As a natural pedestal, bamboo waves rise, revealing bamboo utensils.

第 74-75 页：展览开场——自然之怀。
本页：展览的第一观，竹浪涌向竹林。对页：自然的承台，竹浪上浮现竹器

This page, Opposite, above: As a natural pedestal, bamboo waves rise, revealing bamboo utensils. This page, Opposite, below: The prelude of spatial triptych unfolds—a long scroll of Bamboo Forest, Bamboo Waves, and Painted Boat.

本页，对页，上：自然的承台，竹浪上浮现竹器。本页，对页，下：序幕空间的三段序列长卷——竹林·竹浪·画舫

这是为"东方竹"展览而做的序幕空间,以竹林与竹浪作为幻梦的开局,以情境的方式让观众直接浸润主题,将竹子的东方性直接物化为可以进入的园林场景。

竹林,是中国的自然化的殿堂,是文人的雅集圣地。竹浪,是将管箫之声的形式化与游园化,造就了一片通感之境遇。竹浪也是一种自然化的展台,为展品提供特殊的背景,作为展示的旁白与气氛。

这是当代中国园林进入博物馆展览营造的一次实验,将山水带入了室内,放大了展览的体验方式与观想视野。同时,这次营造活动也是中国美院建筑学院的建筑体验课程,是面向社会开放的以自然为主题、以自然方式营造的艺术与教育活动。

Credits and Data
Project Asia Bamboo Life and Art Exhibition Location: Hangzhou, China
Client: China Academy of Art
Completion: 2018
Principle Designer: WANG Xin, SUN Yu
Construction: Hangzhou Rixing Furniture Company, China Academy of Art student volunteers

This page: Views of the Painted Boat at the top of the bamboo waves. Opposite, left: A vertical scroll of three scenes in the opening space: Bamboo Forest, Bamboo Waves, and Painted Boat. Opposite, right: Volunteers participate in the making of the bamboo waves.

本页：浪顶画舫对竹浪两侧的取景。对页，左：序幕空间的三段序列立轴——竹林·竹浪·画舫。对页，右：志愿者参与竹浪的营造。

SURE Design
Beyond Graphic
Shenzhen, China 2019

烁设计
超越平面
中国，深圳 2019

Comics are a form of art that uses drawings to portray life or current affairs in a simple and exaggerated way, with a strong narrative. Over the past century, the expression of comics has evolved to enhance narrative convenience, vividness, and entertainment by adding exaggeration and graphic expression. With the increasing popularity of comics, graphic languages have become an independent image symbol, widely used in various visual expressions.

We extract the expressive language of comics, such as storyboard, dialog box, speed line, and onomatopoeia, to reshape the picture with a combination of visual languages. By filling in IP images in the exhibition and combining these visual languages, the plots without comic characters are established, compatible with environmental graphics, printed materials, and derivatives, creating a unified visual language for visitors to feel the combination of themselves and the comic scenes while browsing the exhibits, making the exhibition experience more vivid.

pp.82-85: *Y-COMIC-X exhibition, Shenzhen.*

第 82-85 页:"百年国漫大展"展览现场。

漫画是具有较强叙述性的艺术载体，是用简单而夸张的手法来描绘生活或时事的图画。在漫画近百年的发展中，为了增强其叙事的便捷和生动性，或增添夸张、娱乐性和图像表现力，漫画的表达手法也在逐步进化。到了当代，本属于漫画这一亚文化所特有的图像语言，随着漫画文化的普及，已经成为独立的图像符号，而被广泛地运用到各类视觉表达中。

我们将漫画中的表现性语言，如分镜框、对话框、速度线、拟声词等抽离出来，用这些视觉语言的组合重新塑造画面，一方面填入展览中的 IP 形象，使多风格、多角色的漫画能够在现场呈现出统一的视觉语言；另一方面通过这些视觉语言之间的组合，建立没有漫画角色的剧情，与环境图形、印刷物料、衍生品结合相容。参观者在浏览展品的同时，感到自己与漫画场景的结合，使得观展体验更加生动。

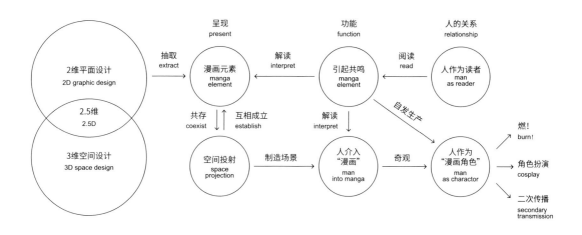

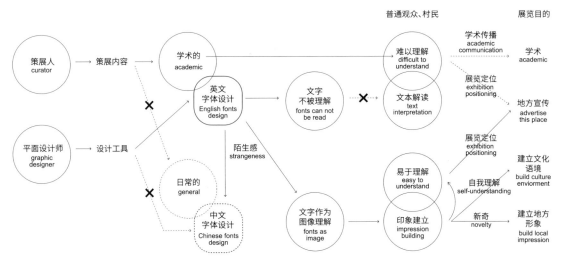

The "Jingkou Revive!" was the theme of the 2017 Bi-city Biennale of Urbanism\Architecture (UABB) Guangming sub-exhibition space. The Jingkou Community located on the edge of the city is a green ecological pastoral community surrounded by agricultural landscapes like farmland, orchard, and reservoirs. This exhibition discussed how to revive Jingkou by developing local cultural production and pastoral life.

Based on the curatorial concept, we designed a set of English fonts as image elements and added graphic textures with local characteristics to the application of the poster. By doing so, the exhibition was endowed with a localized linguistic foundation visually, serving as an international discussion platform. Moreover, the use of English and intense colors reinforced the peculiarity, uniqueness, and contrast of the "international exhibition" to the life of villages and towns by creating a strong contrast and collision in the levels of vision, culture, and context between "enclave embedding" and "localized rural vision."

"迳口复兴！"展览场地位于深圳市光明新区迳口社区，展期为2017年12月23日至2018年2月4日，由深圳青年建筑师尹毓俊策展。光明分展所在的迳口社区周边被农田、果园、水库等农业景观所覆盖，是一个位于城市边缘的绿色生态田园社区，因此本次展览重点讨论的是如何通过在地文化生产和对当地田园生活的发掘来进行城村激活。

我们根据策展理念，为展览设计了一套以英文字体为主的形象元素，在海报的应用中填充了具有迳口社区本土特色的图形肌理。以视觉的方式使展览一方面具有了本土化的语言基础，另一方面也具有国际化讨论平台的性质。同时，展览主视觉通过英文与强烈色彩的应用，突显展览中"飞地性植入"与"本地化乡村视觉"在视觉、文化和语境上的强烈反差和冲撞，再次强化了"国际化展览"对村镇生活带来的独特性、唯一性和反差性。

This page, above, Opposite: Jingkou Revive! exhibition, Shenzhen. This page, below: Jingkou Revive! exhibition catalogue, posters, and products.

本页，上，对页："迳口复兴！"展览现场。本页，下："迳口复兴！"展前册、展览海报及展览周边产品。

Technology and Beyond
不止科技

Intelligent Construction—Future of Architecture
智能建造——建筑的未来
RoboticPlus.AI 大界机器人 (pp. 102–119)

House of Cores 核心之屋
HANNAH (pp. 120–125)

Ashen Cabin 梣木小屋
HANNAH (pp. 126–133)

Qingdao Ruyi Lake Complex 青岛如意湖
Onesight Technology 以见科技 (pp. 134–137)

Casablanca Biennale —"Chopsticks" 卡萨布兰卡艺术双年展——"筷子"
FABO "数制"工坊 (pp. 138–147)

孟浩 MENG HAO
陆唯佳 LESLIE LOK
罗锋 LUO FENG
丁峻峰 DING JUNFENG

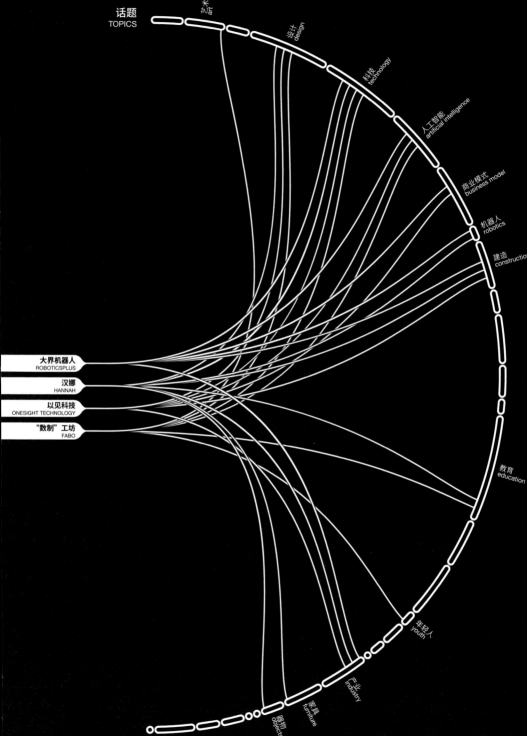

Dialogue:
Technology and Beyond
FAN Ling / MENG Hao / PENG Wu / SU Qi / QIU Wenhao

对话：
不止科技
范凌 / 孟浩 / 彭武 / 苏奇 / 邱文浩

WANG Fei: Today's society has rapidly changed due to technology, which has had a profound and far-reaching impact on every aspect of our life. As young professionals trained in architecture, what do you think is the relationship between architecture and technology? In other words, what are the influences of technology, the internet, and big data on the architectural discipline and industry? How should traditional architecture respond to these influences?

FAN Ling (Tezign) and PENG Wu (Glodon), what motivated you to enter the technology and internet industry after rigorous traditional education in architecture and design and many years of architectural practice?

FAN Ling: I am a maker at heart and enjoy creating things, so studying architecture suits me well, whether it is theory or architectural practice. However, I am not a person with patience. Although drawing and model making are enjoyable, it takes a long time to build a house. I am also concerned about whether a building that consumes so many social resources can be confidently designed in one go. There has probably never been a better time for makers to create something that people can use and receive feedback from users, rather than just making models or drawings.

I enjoy doing things related to architecture and have many architecture books on my bookshelf. However, I do not have the patience for the problems that come with practicing architecture. Starting an internet business allows for more efficient feedback and more frequent contact with society, enabling makers to continuously improve through constant iteration. Such an opportunity did not exist in the past.

I may be different from the other people here, as I am not proficient in coding. I did not have a good opportunity to start a technology business in the 2000s, but in 2010, there was an opportunity for model innovation using technology. Many entrepreneurs in this era have lowered the threshold for technology, as well as the barrier to capital, making it easier for makers to complete something from start to finish.

Of course, all this support comes at a price, such as being responsible for a company with over 400 people, especially in times of economic pressure. I have to learn about the organization, mission, vision, values, organizational culture, KPIs, and other factors to accomplish tasks. My job also involves analyzing financial data and performance. Although I am running a technology company, in fact, my approach to dealing with people is not much more advanced, which may be the price I pay for doing something.

PENG Wu: In my regular routine, I am aware that the way things are done in daily practice differs from what you typically learn in education or in a system. Design always occurs rapidly and problems arise, and you must quickly find a solution in that rhythm. This is something that cannot be easily taught by others, and you must often combine the most recent tools and methodologies yourself to find innovative solutions. During my time working on the architectural form and facade design of Shanghai Center, grasshopper, a visual programming tool, was widely used. Logical and

王飞：当今的社会日新月异，科技对各个方面的影响极为深刻和广泛，作为受到严格建筑教育的年轻一代，你们觉得建筑学与科技的关系是什么？换句话说，科技、互联网、大数据对建筑学及产业有什么样的影响？而传统建筑学应当如何应对这样的影响？

范凌（特赞）和彭武（广联达），什么原因促使你们经受了严格传统的建筑学设计学教育，也做了很多年的建筑实践，却投身到科技、互联网的产业当中？

范凌：我内心就是一个创客，想做东西，建筑学很适合我，不论是建筑理论还是建筑实践，我都学得很开心、很享受。另一方面，我不是一个有耐心的人，我想把东西做出来，但是画图做模型不足以满足我，盖个房子时间很漫长。我也很担心消耗那么多社会资源的建筑，能不能一蹴而就地"自信地"设计出来？有可能从来没有一个时代像现在这样给创客这么好的机会，不是在做模型或者图纸，而是真的可以手把手地建出来。这些东西能给人们使用，使用者也可以给我反馈。

建筑学本身我非常喜欢，我的书架上很多建筑学的书。对于建筑实践所要面对的问题，我的耐心是不够的。创业做互联网可以获得更高效的反馈，会和社会有更高频率的接触，创客可以通过不停的迭代做得更好，过去历史上没有这样的机会。

我与在座的其他几位可能不一样，我的代码写得很烂。从2000年到2010年那一代的技术创业，我没有很好的机会。但在2010年有了通过使用技术而进行模式创新的机会。这个时代有很多创业者把技术的门槛降低了，资本获取的门槛也降低了，对创客来说，从头到尾做完一件事的门槛降低了。

当然，获得这些支持，也必然要付出代价，比如说

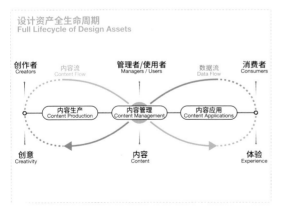

Full Lifecycle of Design Assets © Tezign
设计资产全生命周期 © 特赞

Product Matrix of Tezign © Tezign
特赞的产品矩阵 © 特赞

data-driven reasoning was employed to solve complex design problems. This approach was vastly dissimilar to the conventional model of doing things. However, this new way of working resonated with many younger individuals and co-workers, and it was quickly popularized and expanded throughout architecture schools and firms worldwide. Some of the automated scripts I created were later compiled and utilized by a large number of individuals for their own work. For example, they could generate a large number of building plane diagrams with a single click or conduct correlation analyses between building shapes and housing rates. My script encapsulated the entire process, from geometry to data, and was quickly communicated to everyone.

When I arrived to work in the United States in 2016, I observed a more complete ecosystem in architecture. In American society, the architecture, system, graphics platform, and software you use are all an integral part of a dense forest, with technology at its core. Practical education cannot determine the significance of this matter, and all design firms may face this issue in the future. To increase the value of design, it is important to investigate and expand into the technological realm. The energy, depth, and transformation that design technology will bring are just the beginning, and action in this direction is still valuable and significant.

My own design work primarily relates to technology companies and internet behemoths. Essentially, the demand for real estate throughout the entire San Francisco Bay Area is driven by these technology companies. When interacting with these technology companies, you can sense the entrepreneurial spirit that fosters innovation and competition. This is the fertile ground from which ideas and technologies emerge. In Silicon Valley, you can sense the fusion of design and technology, pragmatism, and insight into the frontier. Digital technology has given these companies strong organizational power, which promotes the value of technological innovation even further. Currently, the total capital density of Silicon Valley will exceed that of a number of medium-sized nations, demonstrating the power of technology.

The United States has a system of architect responsibility, where architects must possess a high level of professionalism and proficiency, and information circulates rapidly. Design technology can be converted into new levels of productivity in very little time, and some exceptional design firms are fundamentally technology companies. These companies have a high level of business digitization, and their design and technology comprise a self-sustaining ecosystem. This development trend is also present in China, where virtually everyone discusses digital transformation and the future form. Innovation and technology will drive future development.

I believe what I am doing right now is appropriate. Traditional architectural education and the use of technology will not conflict and will be closely related in the future.

WANG Fei: MENG Hao (RoboticPlus), SU Qi (Kujiale), and QIU Wenhao (iFLYTEK), how do you reflect on architecture as a discipline, given that your work involves influencing and reshaping it through technology?

MENG Hao: As a maker, I am constantly thinking about how to turn my ideas into reality. I obtained my bachelor's degree in architecture at Melbourne University and I have always been interested in how architecture can be industrialized and realized as a definable product. My focus has been on construction robots and industrial software applications. In future, architects will have a better understanding of intelligent production and will be able to use 3D models of buildings to drive the flexible production of robots to meet the needs of mass-customized projects. My master's and PhD research aim to create a more intelligent industrial software platform for the architecture industry.

I transitioned from architecture to civil engineering in order to study technology and structure, production methods, and to connect data from design to production, which broke down architectural boundaries. In terms of education, I believe that architecture should teach everyone how to be an architect or product manager, and how to solve product, engineering, and software

需对公司 400 多人负责，尤其是现在经济压力大的时候，为了做出一点事还得学习组织、使命、愿景、价值观、组织文化、KPI 等等。我的工作还要看财务数据、做绩效。虽然在做一个科技公司，但事实上和人打交道的方式也没多先进，这也是我为了做些事情所付出的巨大代价。

彭武： 在平时的工作中，我也能感受到在日常实践中做事情的方式和你所接受的教育或者通常所认为的那套体系不太一样。设计总是很快发生，问题接踵而至，你在那种节奏下必须很快找到一个解决方法，而这些解决方法并没有人能教你，需要自己去组合最新的工具或方法论，见招拆招。当时在上海中心的建筑形体和表皮设计上大量使用 grasshopper 这种视觉化的编程工具，用逻辑思维、数据思维来处理复杂的设计问题，设计模型的生成、制作（大量的 3D 打印）和推敲都和传统的模式有很大的不同。十几年前这种做事情的方式迅速引起了周围的一些年轻人或同行的共鸣，大家都愿意来了解你采用这种东西有什么特别的地方，也都在尝试和运用。实际上略有先后吧，这些新的技术和方法体系在全世界的建筑院校和建筑事务所里普及、拓展，蔚然成风。后来我自己编的一些自动化脚本，被很多人收藏拿去做自己的工作，比如一键生成大量建筑平面图解，进行建筑形体与得房率的相关性分析。我写的脚本把从几何到数据的流程全部打包好，问题针对明确，简单易用，因此迅速地被大家拿去分享。

2016 年到美国工作，让我看到一个更完整的生态，在那个社会里面，建筑学、系统、图形平台、编写的软件，所有的东西它都是一个完整的生态，它像一片郁郁葱葱的森林，在这片森林里面最关键的是它的科技属性，你能看到它隐含的强大能量，我们的实践教育里面并没有把这个事情的价值提取出来。实际上，所有的设计公司未来可能都面临这个问题，他们要往科技的方向去探索、去延伸，提高设计的价值。设计科技的那种能量、深度，还有它将会带来的改变，我觉得现在还只是刚刚开始，往这个方向去做一些事情还是很有价值的，会很有意义。

我的本职设计工作，也基本和科技公司、互联网大厂有关，实际上整个旧金山湾区的地产需求基本都是这些科技公司驱动的。和这些科技公司的接触，你能感受到活跃的创业氛围，鼓励创新和竞争，这些是思想和技术所产生的土壤。在硅谷能感受到设计与科技的融合、务实精神、对前沿的洞察，而且数字技术赋予了这些公司强大的组织力，进一步促进了技术创新的价值。目前，整个硅谷总资本的密度会超过一些中等体量的国家，这个就是技术的力量。美国是建筑师负责制，建筑师职业化程度很高，对技能的要求也很高，信息流动很快，设计科技转化为新的生产力的路径很短，一些出色的设计公司本质上就是一家科技公司，其本身的业务数字化程度就很高，它的设计和科技就是一个完整的生态。现在，你能在国内看到，我们也在呈现这样一种发展趋势，几乎所有人都在讨论数字化转型，都在讨论未来形态。技术和创新成为下一步发展的引领，我觉得自己现在做的事情和建筑师的传统教育其实也没有冲突，和未来甚至是高度相关的。

王飞： 孟浩（大界机器人）、苏奇（酷家乐）、邱文浩（科大讯飞）：你们所做的都是通过科技来影响、重塑建筑产业，那么对于建筑学科本身有什么样的反思？

孟浩： 我其实也是个创客，想着怎么样把自己的想法实现。我本科学的是建筑学，在墨尔本大学，因为我是理科思维，当时建筑有参数化技术，我就想建筑是否可以当做可被定义的产品去实现，比如说从设计到建造的数据能不能打通，用工业化的方式生产。我聚焦在建筑机器人和工业软件应用这个方向，希望未来可以让建筑师天马行空的想法有更好的智能生产方式，能够用建筑的三维模型驱动机器人的柔性生产，满足大规模定制化的

Glodon Digital Design BG Design Innovation Center © Glodon
广联达数字设计 BG 创新设计中心 © 广联达

problems, as well as how to integrate civil engineering and architecture together to meet the interests of digitalization, industrialization, and intelligence.

I believe that the future of architecture will involve systematic thinking with an emphasis on vertical depth. For example, the design and production algorithms are interconnected, and engineering knowledge can be better integrated across borders. When I was studying architecture, I wanted to learn programming, but it was not offered in architecture courses. Later, I had the opportunity to go to the Royal Melbourne Institute of Technology's design algorithm laboratory to learn programming.

Therefore, I wonder if we can expand the breadth and depth of architecture-related disciplines in future education and offer students more interdisciplinary courses at an earlier stage.

SU Qi: Our position is quite simple - to provide tools for everyone which provide value for every user. Whether we have contributed value to the industry is more evident when each of our users employs our application. If our users continue to utilize our products, it demonstrates our value. We cannot assert that we have created a disruptive change in the industry after so many years in business. From our perspective, ensuring that each user perceives the value of our tool is the root cause for users to stay and encourage others to use it. When more people use it, it will have the same widespread impact as the modern internet.

Regarding my understanding of architecture, I used to be a teacher in the past and my reflection is that architectural education is somewhat out of date or lagging behind the times. There was a high demand for master-level design because the architectural education I received at the time was in line with the late phase of industrialized production when supply was less than demand. This is the brand effect. In other words, when there is enough work to choose from, a man can focus on something else.

But times are changing, especially in America. When I graduated in 2009 in the United States, the economic crisis just began. At that time, the majority of US companies moved their large-scale projects to China or the Middle East, and there were still regional dividends. However, now the world's dividends have diminished, and there are too many suppliers but less demand for unconventional design, such as the pursuit of master design in the past. When the market changes, I don't believe that education adapts to it. The good news is that I have observed some conscious direction emanating from the Pritzker Prize in recent years. There may be more concern about the design for ordinary people or the value of micro users themselves, and I agree with this because it is more similar to the logic of the internet, which pays more attention to the microscale that can be abstracted.

Finally, I will not engage in architecture because I do not particularly agree with my previous master's level education, which is the primary reason for my reflection. As a result, I ceased instructing at USC and switched to computing at Harvard University.

QIU Wenhao: I will discuss the transition from traditional architecture industry to an artificial intelligence technology company. After making this transition, my most immediate impression is that the architecture industry invests less in R&D than the technology industry, leading to excessively high dividend rates. When labor hours cannot be extended, technology can increase productivity. Therefore, I believe that the industry can collaborate with technology to increase efficiency, allowing humans to focus on tasks that require human decision-making, while machines perform tasks best suited for them. Machines can replace humans at higher levels through continuous training.

In fact, the organization and structure of architects in the industry are similar to the structure of pyramids. Senior architects form a base class of entry-level architects when understood in terms of object-oriented programming. An architect's base class is equivalent to its higher level.

There are three types of work in architecture. The first type involves evaluating and correlating certain rules. This type represents the scope of problems that grasshopper parameterization can

项目需求。我硕士包括博士做的研究，是在建筑领域开发更加智能的工业软件平台。

我从建筑学跨界到土木工程，去研究工艺和结构，研究生产方式、设计到生产的数据串联，这突破了建筑学的设计的边界。回到教育上，我觉得建筑学应该更好地教会大家怎样去做建筑师，做一个架构师，或者去做一个产品经理，解决产品的问题，工程的问题，软件的问题，将土木工程和建筑设计融合，拥抱数字化、工业化和智能化。

我觉得建筑学未来一定是系统化的思维，聚焦纵向的深度，比如说设计和生产的算法打通，工程知识可以更加跨界融合。当时我在建筑专业里面想去学编程，但建筑学的课程并不能覆盖，后面我去了墨尔本皇家理工大学的设计算法实验室，才有这个机会。

所以我在想是不是能够早一点，在未来教育中围绕着建筑学拓展学科的宽度和深度，给学生提供更多的跨学科的课程。

苏奇： 我觉得我们的定位很简单，就是为大家提供工具，所以这个工具必须为每一个使用它的人提供价值。关于我们是不是为行业提供了价值，我觉得这个更多是从我们每一个用户使用这个工具的时候体现出来的。所以如果我们的用户继续使用我们的产品，我觉得这就体现出我们的价值了。我们不敢说对这行业有一个什么颠覆性的变化，我觉得创业这么多年，这种话说出来都太虚了。从我们的角度来说，最根源的就是看到每一个用户在使用这个工具的时候，能够感受到它的价值，然后他会留在这儿，才会有更多的人来用。用的人多了之后，就会产生当代互联网的这种规模化的效应。

如果要说到对于建筑行业的一个理解上，我个人以前也做过一段时间老师，我个人对建筑教育的反思是觉得建筑教育是有点脱离或者滞后于时代的。我那个年代接受的建筑教育，可能更多处于工业化大生产后期，属于供小于求，所以滋生出很多需要大师级设计的需求，或者说很多的人在追求大师的设计方式，本质上就是一个品牌效应。说白了就是当大家不愁活的时候，他就可以想点别的了，就这么简单一个道理。

但是我觉得这个时代在变，尤其是在美国。我 2009 年在美国毕业的时候，正好赶上他们的经济危机，那个时候基本上美国所有的事务所都把他们大规模的项目迁到中国或者中东，当时还有这个地域红利。但是现在来说，全世界的红利也没有那么多了。我觉得现在反而是供需关系上有一个反过来的变化，就是供方太多，但是像之前那种对于大师设计的追求，对于这种标新立异设计的追求的需求变少了。当市场发生了变化的时候，我觉得教育本身并没有适应它。不过，好的一点是从这几年普利茨克奖上，我能看出来他们有意识的一些引导。可能更多关注于给普通人的设计或者是更微观的使用者本身的价值。我觉得这个点我个人是比较认同的，因为他跟互联网的逻辑是比较像的，他更关注于可以抽象出来的微观层面。

所以我的反思可能更多是因为我对于以前的大师式的教育不是特别的认同，这是我最终不会去从事建筑的一个核心原因，也是为什么我不再去南加州大学继续教书，而在哈佛大学的时候转行到计算机方面的原因。

邱文浩： 我就从建筑行业这样一个传统行业到人工智能科技公司这个角度来谈一下，因为我转过去之后，我最直接的感觉是跟科技行业相比，整个建筑行业的研发投入的比例太低，这个背后是分红率太高。在劳动时间无法再延长的情况下，应该通过科技使效率增量。那么通过与科技的合作，我觉得这种行业其实完全可以提高效率，让真正需要人来决策的事情让人来做，让机器能做的事情交给机器做，并且可以通过不断训练人工智能，让机器能够替代的人的等级越来越高。

其实，我们知道在这个行业里，建筑师的组织结构和金字塔也是一样的。那么我们如果以编程里面向对象的方式去理解的话，实际上高级建筑师是低级建筑师的一个基类，就相当于它的一个更高的层级。那么，在目前建筑学的工作中：

第一种工作，是一些规则明确的判断与关联的工作，这是目前 grasshopper 参数化能解决的程度，这也是目前大多数设计院对于新技术的应用水平。

第二种工作，需要对于图纸模型目标进行识别，并且做一些评价标准明确的工作，那么这些工作就可以被 CV 计算机视觉 + 算法替代。

第三种工作，可能评价标准不那么明确，可能还带有很多建筑学的感性与美学在里面，一般这些工作由上

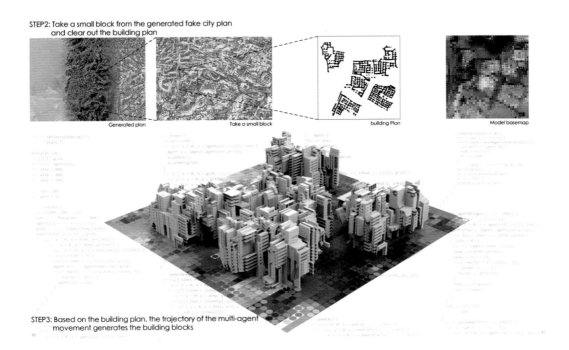

Based on the building plan, the trajectory of multi-agent movement generates the building blocks © Qiu Wenhao
基于人工智能生成的建筑平面，利用多智能体算法生成建筑体块 © 邱文浩

solve and the application level of most design institutes for new technologies.

The second type involves identifying the target of the drawing model and performing tasks with clear evaluation criteria, which can be performed by computer vision (CV) algorithms.

The third type of work requires less distinct evaluation criteria but relies on architectural sensibility and aesthetic awareness. Superior architects typically delegate these responsibilities to subordinate architects. Although the path for completing a task may not be clear, superior architects can easily determine if subordinate architects are performing it well. If different subordinate architects perform the task, the results may not be significantly different, so this task can be performed by AI. What is required is the development of a suitable algorithm. A virtual person is a maximum likelihood estimation of a real person and can guess the most likely judgment made by a real person. From the perspective of senior architects, it is impossible to determine whether the task is performed by humans or artificial intelligence. This is the present circumstance, and it is also the direction in which we hope to conduct research as a technology company specializing in artificial intelligence.

I recommend introducing new technologies in education to help students develop interdisciplinary skills in mathematics and computer technology, so that the best students do not have to just focus on the projects as they do currently, but focus more on examining multidisciplinary exchanges outside of architecture in order to see the current wave of progress.

WANG Fei: Do you have any projections and forecasts for the direction of technology in the future? Such as big data, virtual media, NFT, Metaverse, etc.?

FAN Ling: As an architect, I believe that

STEP 5: Remap generated elevations back to the blank elevations. Transfer the blocks to the Style Architecture.

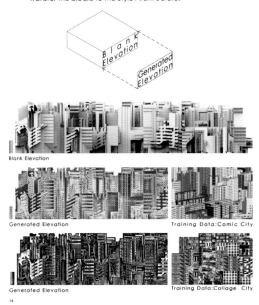

Based on the building plan, the trajectory of multi-agent movement generates the building blocks © Qiu Wenhao
基于人工智能生成的建筑平面，利用多智能体算法生成建筑体块 © 邱文浩

级建筑师分配给下级建筑师做，尽管没有明确的做事的路径，但上级建筑师能够清晰地评价下级建筑师做得好不好，而且换不同的下级建筑师来做，结果可能相差不大，那么这个就是可以被人工智能替代的。我们只要构造出一个虚拟人，对于真人做最大模仿，能够猜出真人最可能做出的判断，那么在上级建筑师看来，其实分不清到底这件事情是人做的还是人工智能做的。这目前也是我们讯飞作为人工智能科技公司希望研究的方向。

我觉得需要引入新技术教育，培养学生一些新的跨学科的技能，尤其是数学，以及计算机技术领域，让优秀的学生能够不只是像现在这样特别关注当下的计划，而能够把更多的精力留在看看建筑以外的多学科的交流上，这样才能看到时代进步的浪潮。

王飞：请问大家对科技的未来走向有什么展望和预测？比如大数据、虚拟媒体、NFT、元宇宙等。

范凌：建筑师是一个知识工作者，知识工作者应该很有杠杆效应，就像写代码的工程师一样，但在建筑师的实践里边，这种杠杆效应有越来越小的趋势。怎么做一个知识工作者的工作空间，这可能是我们会去想的，这个空间也不完全是一个室内空间。除了室内空间这些建筑师可以控制的空间之外，和城市空间会产生一种很好的工作关系。

上海可能是第一个有 5000 台机器组成城市大脑的城市。全世界没有第二个城市有这种运算能力，这些机器的摄像头可以做实时的运算。但再多的人工智能大数据，还是没有办法把城市治理得更好，这里边是有很多提升的机会，可能会需要一个物理空间的设计师和一个虚拟空间的设计师的协作。这个问题可能在中国尤其明显，我们现在一直是要发展的，我也是看好发展的，但如果我们每个人的物质消耗，比如我们每个人住的房子要跟美国人一样大，那把全中国的土地都盖上房子也不够，又比如我们每个人对于肉和水资源的消耗量跟美国

knowledge workers should be highly leveraged, just like engineers who write code. However, in the practice of architecture, this leveraging effect tends to become less and less. We need to think about how to create a working space for knowledge workers, not just an interior space. Architects can also create a good working relationship with urban spaces that they do not control.

Shanghai is probably the first city with 5,000 machines. No other city in the world has this kind of computing power, and the cameras of these machines can perform real-time computing. However, even with all the AI and big data available, there is still no way to govern the city better. There is much room for improvement here, and it may require the collaboration of a designer of physical space and a designer of virtual space. This problem may be particularly obvious in China, where we are always aiming for development. However, the material consumption per capita would be insufficient to provide the lifestyles that Americans have, such as big houses and the consumption of meat and water. This is not an economic issue, but a matter of resource shortage.

Efficient allocation of resources is a technical problem that can make our lives better. Cities make life better, and architects design urban spaces, both indoor and outdoor. However, architects can play a more important role in making our lives better. Does architecture make us safer? Does the primitive hut protect us? Does technology make us more empowered? At this point, neither side has given a good answer yet. How can we reduce carbon and water consumption with more limited resources? This is a technical problem that can only be solved by technology; otherwise, the solution might be war and exploitation, as used by previous generations. In this dimension, it may be irrelevant whether we have a luxury standard office building or not.

PENG Wu: Recently, the issue of Google's AI has gained a lot of attention. Is it true that computing power has progressed to this point? Google's latest big news is the completion of a new office designed by BIG, which is encouraging competition for office space in Silicon Valley. In 2016, I witnessed the first competition at the new Nvidia campus, which was created in Gensler's San Francisco headquarters. The design team next door transformed the entire Nvidia into an airport with a large triangular roof, a completely open office, and an abundance of natural lighting. This time, Google has done an even more thorough job in the third session by employing technology to reduce energy consumption and environmental impact, achieving the ultimate passive design. In the midst of this, there is also Apple's second $5 billion building, the most expensive building ever built by mankind, which is also semi-passive. These facility competitions are extremely active, and it could be the effect of technology companies on the construction industry. It's important to comprehend this trend. If technology can aid in the construction of better buildings, the concept will undoubtedly be maintained, but the ultimate goal should be to benefit people and the environment.

The second person is Ian Goodfellow, the inventor of GANs (Generation Adversarial Networks) in the current metaverse market. He recently resigned from Apple because they require employees to work in the office at least twice a week. Ian finds this unacceptable because he sees no problem with his team working in the online metaverse. So, Ian left Apple and joined Google. Despite Google's fantastic new headquarters, they do not require everyone to work on site, which seems contradictory. It's fascinating how these things happen, and we must sometimes view the impact of technology on us in reverse. We must find new ways to predict what our buildings will become. As FAN Ling stated, we must return to small and more people-centric scenes. In contrast to Apple's icy coldness, Google's new headquarters demonstrates the warmer connectedness that technological advancement should foster, particularly in the context of a metaverse of virtual connections in which individuals are largely alone. I'm reminded of Christopher Alexander's book *Pattern Language*, which emphasizes that the user knows what they want better than the architect and the correlation of various design elements in this context. Alexander's advanced theory in architecture plays a critical role in promoting the development of current technology, focusing on a

一样大，其实就可能不是经济的问题了，是资源做不到。

所以怎么样能更高效地调配资源，同时我们的生活又能变得越来越好，这是一个技术问题。城市让生活更美好，建筑师设计了城市的空间，不管是室内还是室外，但是在使我们的生活更美好这件事情上，建筑师可以发挥更多作用。所以如果科技和建筑的关系回到最原始的一个问题，就是建筑是不是让我们更安全了，原始小屋让我们被保护了，然后科技是不是让我们更加被赋能了？我觉得在这个点上，两边都还没有给很好的答案。在更有限的资源下，如何减少碳的消耗以及水的消耗呢？所以这是个技术问题，或者说这是只有科技才能解决的问题，否则可能会像上一代的解决方式一样，用战争和剥削来进行。在这个维度下可能和我们有没有一个豪华的标准的办公楼是无关的。

彭武： 最近谷歌人工智能确实很受关注，确实算力已经到了这个程度吗？谷歌最近更大的事情是他们请 BIG 设计的新办公室建成了。它的建成进一步推动了硅谷大厂办公设施的全面竞赛。这个竞赛我在 2016 年的时候见证了第一场，是 Nvidia 的新园区，在 Gensler 旧金山办公室设计的，我们隔壁设计团队把整个 Nvidia 设计成了机场风，三角形大屋顶，全开放式办公，大量的自然采光。这次谷歌第三场做得更加彻底了，以技术的方式实现更少的能源消耗，更小的环境影响，被动式做到极致。中间第二场苹果 50 亿美元的人类有史以来最昂贵的建筑，也是半被动式的。设施竞赛发生得非常生猛，这也许就是科技公司对建筑行业的示范效应吧。我觉得这个趋势值得我们去了解，科技如果帮助创造更好的建筑，它毫无疑问肯定要保持它的观念，目标回到人和环境本身。

第二个是现在这种元宇宙市场中 GANs（生成对抗网络）的发明人，伊恩·古德菲洛，他这两天据说从苹果辞职了，因为苹果要求员工回办公室上班，每人每个礼拜至少去两天。古德菲洛觉得不可接受，他认为他的团队线上元宇宙上班没有任何问题。于是古德菲洛从苹果辞职转身去谷歌上班了，谷歌虽然修了这么好的新总部，但是并不强制大家都去，听着挺矛盾的。这些事情发生我觉得比较有意思，有时候科技对于我们的影响需要反过来看，我们现在要用什么样的方式去看我们的建筑要做成什么样，就像范凌说的，要回归那一个小的、更多的跟人有关的场景。相比苹果的冰冷，谷歌的新总部展现了更温暖的关联性，技术的发展应该促进的就是这种关联性，尤其是在个体大量独处的元宇宙虚拟连接的情境下。我想起克里斯托弗·亚历山大写的《模式语言》，它从人的感知出发，强调使用者比建筑师更了解他要什么，以及这种背景下各种设计元素的关联。在 20 世纪 70 年代发表这样的"小微"理论，当时是太超前了，甚至和彼德·埃森曼还有一个很大的争论。亚历山大在建筑领域的超前理论对于当前科技的发展却是有一个决定性的推动作用，关注一个更内在的、更接近于科学的、基于人性的真正的需求，不光是在建筑里面有，这在科技里面也是普遍存在的。

苏奇： 科技界经常说一句话就是"我们曾经想要的是会飞的汽车，结果我们得到的是一个写了 140 字的推特。"我觉得大家不用对科技有太多的一个期望，我们现在看到的这些科技产品可能是大家觉得比较新的东西，但其实硅谷在 20 世纪 60 年代到 70 年代就已经有过探讨和畅想，但直到几十年之后才发生。所以我觉得人类对于科技本身的促进并没有那么快，就像范凌说的，一个突发事情带来的变化，往往比你科技能带来的变化快太多了。

我觉得谈科技或者谈建筑，就不要把自己想得太重要，包括从事互联网。你觉得你的技术能改变世界，但是如果你连一个用户都征服不了，凭什么改变世界。我以前从事建筑的时候，有一个问题我记得非常清楚，在哈佛的时候，我去听过一次史蒂芬·霍尔的演讲，他有一句话特别打动我，他说现在的学生都不会去关注一个人走到建筑里边，看到的光影，感受到空间的变化，相反都在跟我聊不同的形式，不同的主义，他觉得很悲伤，这点我是特别同意的。我觉得大家关注太多的关于社会的东西，或者概念上的东西，而忽略了我们本身。无论是科技也好，建筑也好，都是为每个人提供价值。

如果真是科技对生活产生的价值，我觉得应该都是在很微观的情境中，比方说很简单的，从内容消费的角度，从微信的朋友圈或者公众号到抖音直播以及短视频的转变。当这种类型的产品能够被探索出来，它会对整个产业的生产关系做出很大的改变。往往是这种看起来

more intrinsic need based on true human nature, not only in architecture but also in technology.

SU Qi: In the technology industry, there's a popular saying that goes, "we used to dream of flying cars, but instead we got 140-character tweets." However, I don't think we should have too much expectations for technology. Even though these products may seem new, Silicon Valley has been discussing and imagining them since the 1960s and 1970s, but it took decades for them to become a reality. Therefore, I don't believe that humans can promote technology at such a rapid pace. As FAN Ling mentioned, the changes brought on by an epidemic can often be much faster than those brought on by technology.

When discussing technology or architecture, we shouldn't focus solely on ourselves, including the internet. You may think that your technology can change the world, but how can it do that if you can't even win over a single user? I recall a question that touched me from my time studying architecture at Harvard. During a speech by Steven Holl, he expressed disappointment that students were more interested in discussing different forms and doctrines rather than experiencing the light and space exchange in buildings for themselves. I agreed with him on this. I believe that people place too much emphasis on social or conceptual matters and overlook the fact that technology and architecture should be about providing value for everyone.

If technology truly adds value to people's lives, it should be demonstrated in micro situations, like WeChat Moments or public accounts and live broadcasts, such as Tiktok, from the perspective of content consumption. When such products become commercially viable, they can significantly alter the industry's production dynamics. It's often these seemingly insignificant events that bring about fundamental changes and reforms in our society's production relations, rather than grand concepts that are implemented from the top down and ultimately transform the entire society. Personally, I find that approach implausible. Instead, it's essential to consider the values and needs reflected by each individual, even those who may seem similar. That, to me, is the most important thing of all.

MENG Hao: I wholeheartedly agree with SU Qi. Let's take construction as an example. A construction project involves a long industrial supply chain, with an extensive life cycle, procedures, and parties involved, each with their own professionalism, needs, and objectives. Architects don't like duplicating drawings, while factories want to eliminate labor and streamline production. Additionally, general contractors aim to improve quality and productivity on-site. These are all fundamental human needs that require more intelligent technology to address. It's not about artificial intelligence or high-tech equipment, but about tools that can help professionals enhance their work efficiency and satisfaction. The engineering innovation of life is where these specifics play a role. Therefore, I believe that technology or AI can enhance workflow and be a tool to solve engineering problems based on the small needs of construction industry practitioners. I think this is the fundamental value of technology, regardless of whether it's AI.

QIU Wenhao: I fully agree with MENG Hao's point of view that the purpose of technology is to provide people with new knowledge and tools to improve their lives. Our ultimate goal is always to improve our quality of life. We may transition from traditional office and life models to a more three-dimensional way of connecting and organizing people, things, and things, possibly even in a metaverse prototype or a higher-level version of it. In such a technological transition, we will generate a significant amount of data. As the capacity to process data improves and artificial intelligence continues to advance, we will face this situation for some time. However, the development of hardware, such as processors, has reached a bottleneck, and there won't be any significant progress until new material technology emerges. But in the short to medium term, software still has a lot of room for improvement to match the hardware. For instance, in the metaverse or in fields such as edge computing, it may make our lives and social organization even more distinctive than before, and it is even possible that a new legal ethics-based foundational environment may emerge.

非常小的事情，对我们的社会的生产关系起到了根源性的变化与改革。并不是说一个很大的概念能从上而下地贯彻下来最后改变了整个社会，我个人是觉得不太现实。所以我觉得还是关注到每一个微观的个体，每一个看起来很类似的个体所反映出来的一些价值和需求，我觉得是最关键的。

孟浩： 我非常同意苏奇的一些看法，以建筑为例，回到建筑项目上，这是一个非常长的产业链，一个项目的周期也好，包括它的流程也好，参与方也好，其实非常的复杂，每个参与者都有自己的专业性，有自己的需求和目的。建筑师不想重复地画图，工厂希望可以摆脱人力，进行工业化的生产，总包也想提高现场施工的质量和效率，其实这些都是最人性的需求。这些细节的点需要更聪明的科技来做改变，它不需要人工智能或者是多高大上的科技，它其实是一些能够帮助专业人员去提高工作效率，去享受生活的工程创新。所以我觉得科技或者是人工智能可以授权工作流程，它是个工具，从建筑行业从业者的一些小的需求出发，能够解决一个工程的问题。我觉得这是技术的最核心的价值，不管它是不是人工智能。

邱文浩： 我也同意刚刚孟浩这个观点，技术的目标其实还是以人为本，通过技术、新知识、工具，营造更美好的生活才是最终目的。我们从以前一个比较传统的办公的模式以及生活的模式，可能会进入一个人与人、人与物、物与物之间更加立体的联系以及组织的一个方式。当然这就可能有最近的这样一个元宇宙的雏形，或是元宇宙的更高级的一个形式，可能是一个全方位的连接。

那么紧接着，在这样的一个技术的变革中，实际上我们就会产生很多数据，那么不同结构与维度的数据出现了之后，其实数据处理的能力也在提高，紧接着人工智能也会不断加强。实际上我们应该会面临一段时间的这样的一个情况，因为从硬件的角度上来看，硬件的像处理器的发展应该已经到了一定的瓶颈，在没有全新的材料工艺或者构架出现之前，应该没有太大的进步，这是我们目前中短期能够看到的情况，但是软件的进步还没有饱和，所以中短期未来的软件应该会有更大发展，起码应该能取得能满足硬件需求的进步。然后我甚至还想过，比如说在元宇宙，或者边缘计算等领域，实际上可能会让我们的生活以及社会组织形式与以前更不一样，甚至有可能会出现一个全新的法律伦理道德基础的环境。

RoboticPlus.AI
Intelligent Construction—Future of Architecture
Changsha, chengde, Shanghai, China 2021-2022

天界机器人
智能建造——建筑的未来
中国，长沙，承德，上海 2021-2022

Innovation in Construction Industry

For many years, technological development in the construction industry has lagged behind many other industries. Most of the materials and processes used today are based on knowledge accumulated over the last century. Workers remain the primary contributors to the building construction process, whether on construction sites or in building factories. Meanwhile, in other advanced manufacturing industries, fully automated unmanned factories are already in use. However, due to the complex environment, diverse materials and processes, the contradiction between personalized design and standard installation processes, and low levels of informatization, building construction still relies heavily on human experience for production.

RoboticPlus.AI believes that the construction industry urgently needs technological change. They propose to transfer on-site labor-intensive work to factory automatic prefabrication, transform the production mode that relies on labor into automatic equipment, and convert the operation mode that depends on workers' experience into mass customization using equipment that recognizes components through intelligent algorithms. In the industrial software RoBIM developed by RoboticPlus.AI, various professional disciplines, such as visual algorithms, automatic graphics, and mechanical arm algorithms, are integrated to solve the problems faced by the industrial production of buildings. The software aims to improve construction efficiency and accuracy, optimize construction costs, and ultimately achieve a balance between construction quality, engineering, and economic benefits.

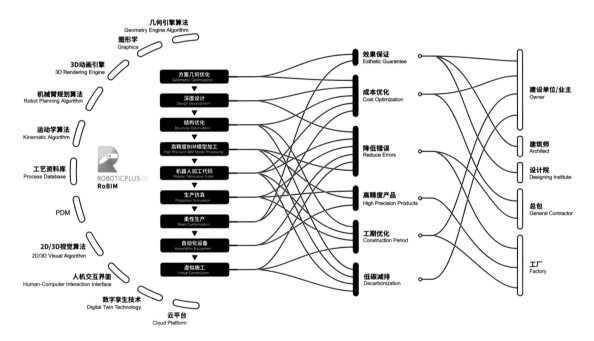

The core tech and value of the industrrial software: RoBIM by RoboticPlus.AI
大界 RoBIM 工业软件核心技术与功能价值

pp.102-103: *Arrival is located in the middle of the Mixc Plaza in Shanghai,* © *CreatAR Images (2022).*
第 102-103 页：降临装置位于上海华润万象城星空广场中央，© CreatAR Images（2022 年）。

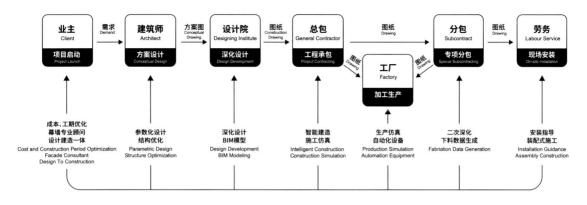

Services provided by RoboticPlus.AI in different building developing stages

大界在建筑开发各阶段的赋能服务

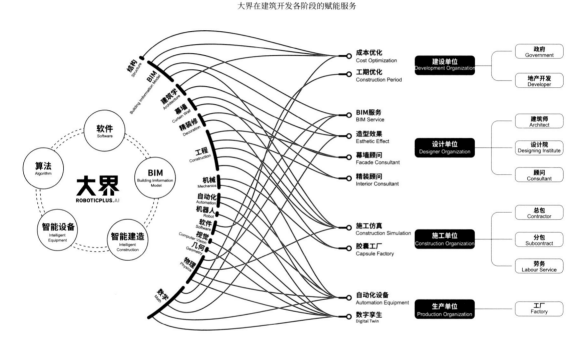

The multidisciplinary team and service provided by RoboticPlus.AI

大界多学科的团队组成与服务内容

建筑建造的变革

长久以来，建筑产业在人类科技发展的历史进程中，与众多产业相比，是最慢的一环。当今使用的材料与工艺大多来源于20世纪的积累与沉淀。不论是在建筑工地还是在建筑产品工厂，人类依然是建筑建造的过程中主要产能的贡献者。而其他先进制造业，全自动化的无人工厂早已落地投入使用，反观建筑建造受限于环境复杂、材料工艺多样、个性化设计与标准安装工艺相悖、信息化程度低等因素，仍大量依靠人工经验来进行生产。

大界认为建筑产业急需一场技术变革，将依赖于现场作业的劳动密集型工作转移到工厂自动化加工预制，将依赖人工产能的生产模式转变为依赖自动化设备生产，将依赖工人经验型作业模式转变为设备通过智能算法识别构件的柔性生产。在大界自研建筑工业软件RoBIM中，综合了各个专业学科，例如视觉算法、自动化图形学和机械臂算法等，解决建筑工业化生产面临的问题，提高建造效率与精度，同时优化建造成本，最终达到建筑效果、建筑工程、经济效益的平衡。

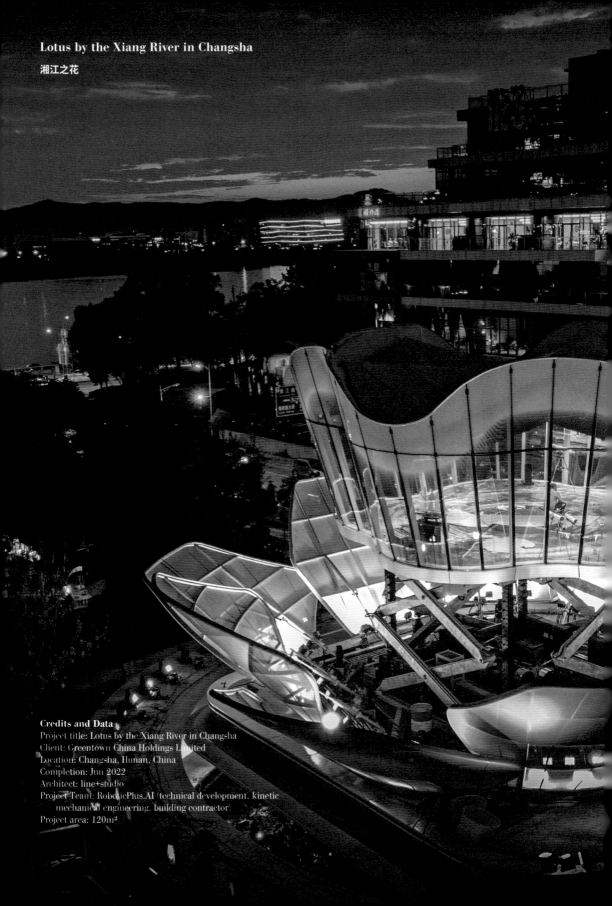

Lotus by the Xiang River in Changsha
湘江之花

Credits and Data
Project title: Lotus by the Xiang River in Changsha
Client: Greentown China Holdings Limited
Location: Changsha, Hunan, China
Completion: Jun 2022
Architect: line+studio
Project Team: RoboticPlus.AI (technical development, kinetic mechanical engineering, building contractor)
Project area: 120m²

The project is situated on the bank of the Xiang River in Changsha and integrates multiple functions like cultural tourism, exhibition, and sales center with movable opening device. The architecture is designed by line+studio. The overall shape imulates the shape of azalea flowers. The structure is composed of stamens (sightseeing hall) and petals. The sightseeing hall can be escalated, lifting visitors from the corridor to the roof layer. As it rises, the petals unfold, and visitors are treated to views of the river scenery.

Playing the role of technical development and building contractor, RoboticPlus bridges the data from the design process to the construction, smart factory production, and prefabricated methods, so that we could stick to the original design during production and construction.

本项目位于长沙湘江畔，是集文旅观光、展览展示、销售中心等多功能于一体的可动式屋面开合装置。建筑方案由line+建筑事务所打造，整体造型模拟杜鹃花朵的体态，由花蕊（观光厅）及花瓣组成，观光厅可向上抬升，带领参观者从廊道跃升至屋面层，上升的同时花瓣展开，江景映入眼帘。

大界在项目中承担设计深化至施工落地的角色，其中设计阶段采用全流程数字化优化，后期通过智能工厂构件生产＋装配式的方式，实现全流程精度可控，最大化还原建筑师方案效果。

This page: the site of Lotus, © WEN Zhang (2022). Opposite: Lotus is located by the Xiang River in Changsha, visitors can view the Xiang River in the interior hall, © WEN Zhang (2022).

本页：湘江之花场地，© 文章（2022年）。对页：湘江之花位于长沙湘江边上，游客可以从内厅观看到江景，© 文章（2022年）。

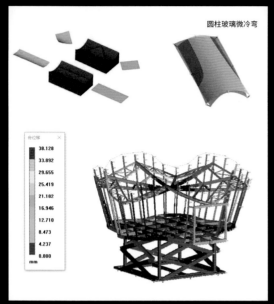
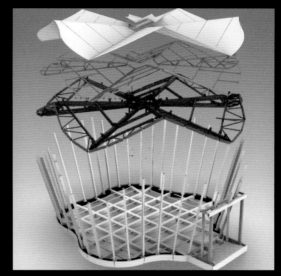
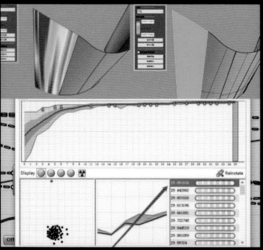
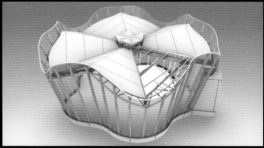
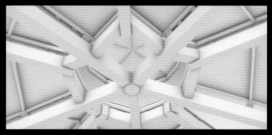
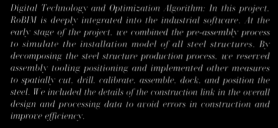

Digital Technology and Optimization Algorithm: In this project, RoBIM is deeply integrated into the industrial software. At the early stage of the project, we combined the pre-assembly process to simulate the installation model of all steel structures. By decomposing the steel structure production process, we reserved assembly tooling positioning and implemented other measures to spatially cut, drill, calibrate, assemble, dock, and position the steel. We included the details of the construction link in the overall design and processing data to avoid errors in construction and improve efficiency.

数字技术与优化算法：项目中大界工业软件 RoBIM 深度融入。在项目前期，我们结合预装配工艺对所有的钢结构进行模型安装仿真。通过钢结构生产工艺进行分解、预留装配工装定位等措施，对钢材进行空间切割、开孔、标定装配对接定位，把施工环节的细节放入整体设计及加工的数据中，避免施工中的误差，提高效率。

Building Information Model for Manufacturing: The BIM model is developed to the precision of LOD400, and it details each component and system, such as the single curved glass system and division, the planking of the roof system, the installation angle, and the connection part of each intersection line of the main steel structure. The model data is directly connected with glass, steel structure, roof and other factories, and the order and material production is tracked from the RoBIM platform. At the same time, the high-precision physical information model can be used to simulate transportation, loading and unloading, storage yard, hoisting, installation sequence, etc., which greatly improves the granularity of project management.

建筑生产信息模型：大界将设计端 BIM 模型深化至 LOD400 的精度，精细到每个系统的每个构件。如幕墙系统的单曲玻璃系统及其分隔缝隙；屋面系统的顶层铺板、安装角度，主体钢结构的每个相贯线衔接部位等。模型成果直接对接玻璃、钢结构、屋面等构件生产工厂，从 RoBIM 平台上进行下单和物料生产追踪。同时，高精度的物理信息模型能够被运用于生产以外的运输、装卸、堆场、吊装、安装顺序等模拟，大大提高了工程项目管理的颗粒度。

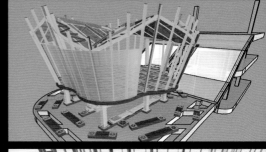

Intelligent Mass Customization: In the construction and production stage, we use the RoBIM platform to integrate all model information, achieve data connectivity between different intelligent processing equipment, and automate the scheduling of NC equipment in the entire production process to achieve model data interaction. During processing, we also used a 7-axis mechanical arm to weld components in the factory to improve production accuracy and efficiency and completed pre-assembly and construction simultaneously. After retesting, the overall error of the overall steel frame assembly was within 3mm.

智能柔性化生产：在建造生产阶段，我们使用 RoBIM 平台整合所有模型信息，实现不同智能加工设备之间的数据打通，整个生产流程的自动化数控设备排单，实现模型数据交互。加工期间我们还使用了 7 轴机械臂在工厂进行构件焊接，提高生产精度及效率，并同步完成预装配搭建，经过复测整体钢架装配误差在 3mm 以内。

Virtual Construction Simulation: Following the pre-assembly stage, and considering transportation limitations, the steel structures will be disassembled into several large components for quick hoisting and installation at the site, enabling the industrialized manufacturing of customized buildings. Faced with time constraints from the owner, RoboticPlus has developed a synchronous construction solution for mechanical devices and steel structures. During the construction simulation of RoBIM, we tackled the installation sequence problem, prepared the necessary mechanical measures in advance, and successfully completed the installation within the designated timeframe according to the simulation plan.

虚拟施工仿真：预组装后，结合运输条件，分拆成若干大型构件装车，抵达现场快速吊装作业，实现定制化建筑的工业化制造。面对业主的工期压力，大界规划了一版机械装置与钢结构同步施工的解决方案，在 RoBIM 的施工仿真过程中，我们解决了安装顺序问题，提前准备需要的机械措施，最终按照仿真计划在工期内完成安装。

Zen Hall of The Upper-Cloister
禅堂

Credits and Data
Project title: Zen Hall of The Upper-Cloister in Golden Mountain
Client: aranya
Location: Chengde, Hebei, China
Completion: December 2021
Architect: Atelier Deshaus
Project Team: AND Office (structural engineer), RoboticPlus.
  AI (technical development, robotic fabrication, building
  contractor)
Project area: 162 m²

The project is located in Jinshan Mountain of Chengde, Hebei Province, amidst mountains, making transportation highly challenging. The project requires a meditation hall to be built within the wild mountain forest. The architectural design was completed by Deshaus Architecture Office in collaboration with AND Structural Office and RoboticPlus.AI, resulting in the first and largest exhibition pavilion with a full carbon fiber roof structure in China.

This page: the site of Lotus, © Zee Leong (2022). Opposite: the interior view and the skylight of Zen Hall, © Zee Leong (2022).

本页：禅堂场地航拍，© 梁喆（2022 年）。对页：禅堂室内与天窗，© 梁喆（2022 年）。

本项目位于河北承德金山岭，地处群山峻岭之中，因此交通极为不便。项目要求是在此野外山林建造一处禅堂。建筑方案设计由大舍建筑事务所完成。这座中国首座也是最大一座使用全碳纤维屋面结构的展亭，整体是由大舍联合和作结构建筑研究所与上海大界机器人公司实现的。

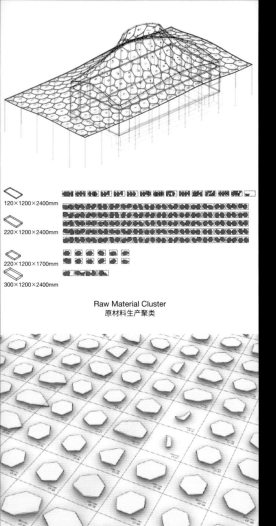

Raw Material Cluster
原材料生产聚类

120×1200×2400mm
220×1200×2400mm
220×1200×1700mm
300×1200×2400mm

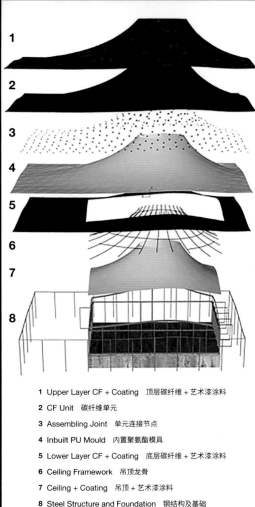

1 Upper Layer CF + Coating   顶层碳纤维 + 艺术漆涂料
2 CF Unit   碳纤维单元
3 Assembling Joint   单元连接节点
4 Inbuilt PU Mould   内置聚氨酯模具
5 Lower Layer CF + Coating   底层碳纤维 + 艺术漆涂料
6 Ceiling Framework   吊顶龙骨
7 Ceiling + Coating   吊顶 + 艺术漆涂料
8 Steel Structure and Foundation   钢结构及基础

Digital Technology and Optimization Algorithm: The carbon fiber materials used in this project are composed of 230 hexagonal modules of varying shapes. In order to ensure the uniformity of processing sizes, specification clustering of raw materials, and other factors, we used dynamic simulation to adjust the 230 modules of different sizes into a version with similar sizes. Through genetic algorithms, we also adjusted the posture of each component in the processing process, aiming to optimize the raw material specifications of the material reduction process from over 100 to four.

数字技术与优化算法：本项目中的碳纤维材料，共有 230 个六边形模块，形态各异，各不相同。考虑到加工尺寸的统一性，原材料的规格聚类等等因素，我们通过动态仿真，将大小不一的 230 个模块调整成为尺寸接近的一版方案，同时通过遗传算法调整每一片构件在加工过程中的姿态，目的在于把减材工艺的原材料规格，由一百多种优化至四种。

Building Information Model for Manufacturing: As the roof installation components in the project are complex in shape, the production cycle is long, the node process is new, and pre assembly is required in the factory. A fine building production information model can connect the design end and the factory end, help the architect to refine the nodes, and also help connect the production plants, so as to jointly develop installation nodes suitable for new carbon fiber materials and new processes, and achieve the purpose of rapid assembly on site. At the same time, it can help the construction end of the project to carefully consider the size distribution of transportation and subpackage, as well as the difficulty of hoisting in the harsh mountain environment when the production of special-shaped components is not completed.

建筑生产信息模型：由于项目中的屋面安装构件形复杂，生产周期较久，节点工艺较新，且需要在工厂进行预组装。一个精细的建筑生产信息模型能够打通设计端与工厂端，帮助建筑师推敲节点，也帮助对接生产工厂，一同研发适合碳纤维新材料新工艺的安装节点，实现现场快速组装的目的。与此同时，能够帮助项目施工端在异形构件生产未完成之际，细致地考虑运输分装的尺寸分配，以及

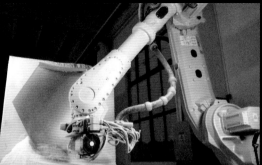

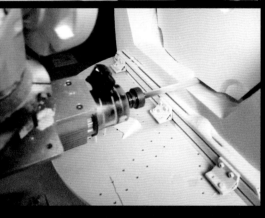

Intelligent Mass Customization: During the manufacturing process, through tests and proofing, combined with foam mold material robotic milling, the freeform mold was realized, enabling the molding embryo of the roof to be produced. Additionally, combining the mature composite material vacuum diversion technology, it broke through the traditional FRP outer mold process and innovatively used the sandwich production method, using the robot to carve and mill the foam inner mold, and producing the roof components through the wrapping, sealing, and vacuum diversion of carbon fiber cloth. The site installation status was restored 1:1 through factory pre-installation.

智能柔性化生产：在制作工艺上，通过多次测试打样，结合机械臂的泡沫模具减材工艺，实现了双曲面造型模具的雕刻铣削成型，使得屋顶的造型胚材实现生产。同时结合成熟的复合材料真空导流技术，突破传统的玻璃钢外模工艺，创新性地使用夹心制作的方式，使用机器人雕刻铣削成型的泡沫内模具，通过碳纤维布的包裹、密封、真空导流，实现屋面构件的生产。并通过工厂预安装的方式，1:1还原现场的安装状态。

Virtual Construction Simulation: After the component products were produced in the factory, they were assembled one-to-one according to the on-site dimensions, and factory acceptance was carried out. Subsequently, 230 components were split into 10 sets of pre-installed modules, using the transportation specification as the splitting logic, greatly reducing on-site construction operation time. Due to the extremely lightweight component weight in the early stage, and the pre-installation, the component was only divided into 10 modules, enabling rapid lifting and installation on-site. It took only one day to hoist the roof structure layer with a developed area of 177 square meters.

虚拟施工仿真：构件产品在工厂完成生产后，依据现场尺寸，1:1完成组装，并进行工厂验收。随后以运输规格作为拆分逻辑，将230个构件拆分成10套预安装模块。大大减少了现场的施工作业时间，对于现场施工方面，由于前期极度地轻量化了构件重量，以及预安装把构件仅仅分为了10个模块，因此现场能够实现快速吊装安装。仅仅用了一天时间，就把177平方米展开面积的屋面结构层吊装完毕。

Arrival
降临

This project represents the highest art installation ever constructed in China using 3D spatial printing additive manufacturing technology. The design scheme was completed by WutopiaLab, while RoboticPlus.AI was responsible for many of the project's professional aspects, including computer parametric deepening design, R&D and testing of the production process, robotic operation programming, and final production installation.

本项目是迄今国内高度最高的，通过三维空间打印增材制造技术建造的艺术装置/构筑物。其方案设计由WutopiaLab完成，大界负责项目中多个专业内容，如计算机参数化深化设计、生产工艺研发测试、机器人作业编程及最后生产安装等。

Credits and Data
Project title: Arrival
Client: China Resources
Location: Shanghai, China
Completion: December 2021
Architect: WutopiaLab
Project Team: RoboticPlus.AI (technical development, 3D printing, lighting, building contractor)
Project area: 9m²

This page: Arrival is located in the middle of the Mixc Plaza in Shanghai, © CreatAR Images (2022). Opposite: the entrance of the installation Arrival, © CreatAR Images (2022).
本页：降临装置位于上海华润万象城星空广场中央，© CreatAR Images（2022年）。对页：降临装置正面入口，© CreatAR Images（2022年）。

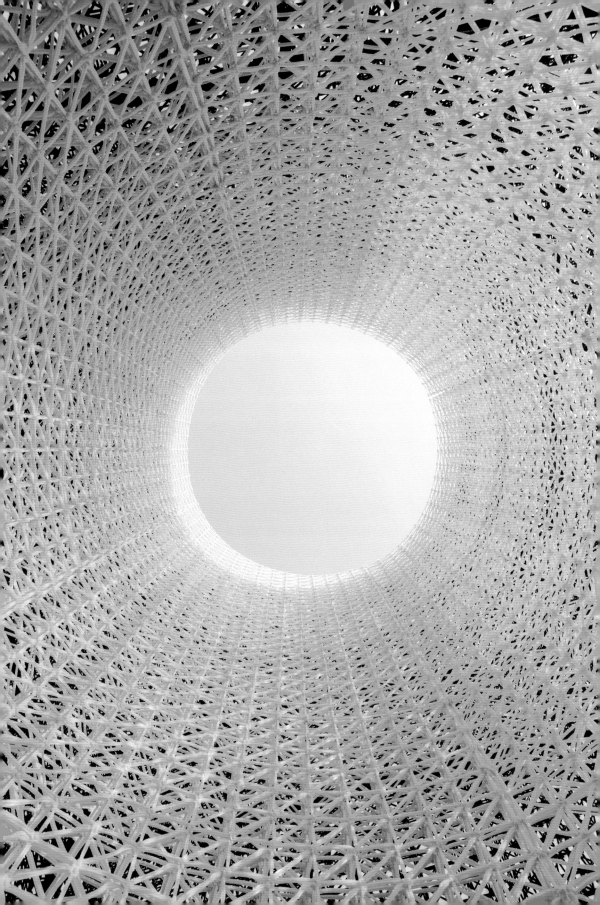

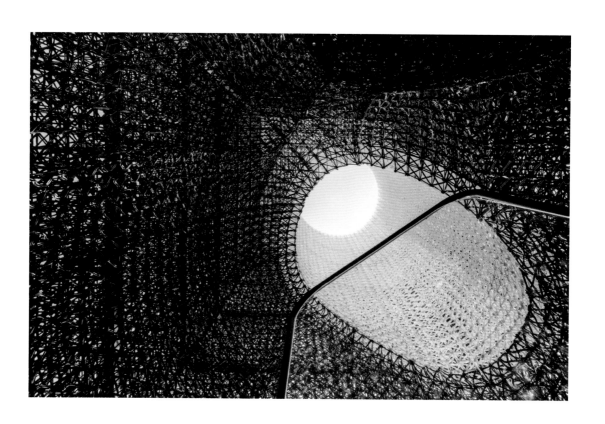

This page: the main material for 3D printing is black (exterior) and yellow (interior), © CreatAR Images (2022). Opposite: visitors can view the sky from the skylight rounded by 3D printing, © CreatAR Images (2022).

本页：进入降临装置内部可发现材料由黑色转变为黄色，© CreatAR Images（2022 年）。对页：进入降临装置内部平台，可通过由 3D 打印构件组成的天井看到天空，© CreatAR Images（2022 年）。

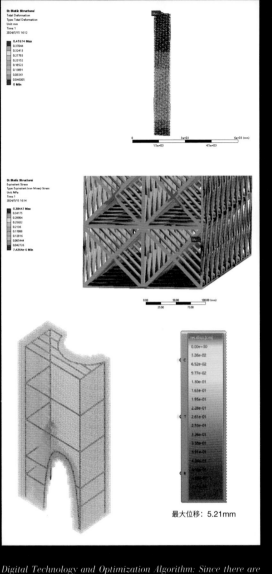
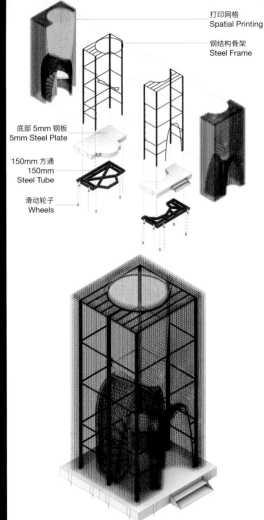

最大位移：5.21mm

Digital Technology and Optimization Algorithm: Since there are few research results on the production process of robotic spatial printing additive in China, the testing and development process of this project was extremely challenging. Through the simulation of the spatial structure, we reasonably optimized the lattice density of the component, and parametrically adjusted the lattice density printed in space from low to high, from dense to loose, according to the component location and force size. The density design of different specifications was directly combined with the processing technology to achieve the unification of the digital design effect and production technology.

数字技术与优化算法：由于机器人空间打印增材制作工艺在国内研究成果不多，本项目的测试与开发过程是极其严峻的考验。我们通过对空间结构的仿真模拟，合理优化构件的晶格密度，把空间打印的晶格密度根据构件位置、受力大小，由低到高、由密到疏松，进行参数化调整。并把不同规格的密度设计直接结合到加工工艺当中，达到数字化设计效果与制作技术统一。

Building Information Model for Manufacturing: The decorative surface of this project mainly consists of decorative components of space printed lattice structure, which is an unconventional building material and difficult to express in two-dimensional drawings. Therefore, a detailed building production information model was created to help architects, curators, and owners understand the final effect of the finished products after production and installation. On the one hand, this formed an effective communication bridge, while on the other hand, it guided the technical test team to effectively discuss and improve the process for solving problems such as installation mode and printing area distribution and provided timely feedback to improve the test R&D efficiency.

建筑生产信息模型：本项目的装饰面主要材料为空间打印晶格结构的装饰构件，是非常规建筑材料，在二维图纸中较难表达，因此建致的建筑生产信息模型在一定程度上帮助建筑师、策展方、业主方去理解最终的生产出来组合安装后的成品效果。一方面形成了有效的沟通桥梁，另一方面可以指导技术测试团队针对解决安装方式和打印区域分布等问题进行有效的探讨和工艺的改良，及时反馈，提高测试研发效率。

Intelligent Mass Customization: In this project, robotic arms were used as production equipment. They were used in conjunction with the 3D printing extrusion gun independently developed by RoboticPlus to form a closed loop for design production. The challenge was to achieve the lightest possible structural mass while meeting the modeling requirements. Despite the fact that plastic additive manufacturing technology in China still widely uses FDM (melt accumulation) molding technology, the RoboticPlus team ventured to use space printing in this actual project.

智能柔性化生产：项目中使用了六轴机械臂作为生产制作设备。配合大界自主研发的三维打印挤塑枪，形成设计—生产—建造的闭环。其中的难点在于，需要在满足造型要求的前提下，得到最轻量化的结构体量；在国内塑料增材制造技术还大范围使用在FDM（熔融堆积）成型技术的情况下，大界团队在这个实际项目中挑战了空间打印。

Virtual Construction Simulation: The installation exhibition site is located in the center of a busy business district, which limits the space and time available for on-site construction and operation. Therefore, the team first analyzed the site situation and developed efficient transportation, unloading, and installation plans based on the detailed building production information model. They carried out construction simulations in a virtual environment to improve on-site construction efficiency. Additionally, they conducted one-to-one pre-construction in the factory space to simulate site conditions and identify the feasibility and difficulty of installation nodes.

虚拟施工仿真：由于装置展览现场位于热闹商圈的广场中央，因此现场施工作业空间、时间有限，必须先对现场情况进行分析，根据精细的建筑生产信息模型，制定高效的运输、卸载、安装方案。进行虚拟环境中的施工仿真，提高现场实施的施工作业效率。同时，在工厂空间进行一比一的预搭建，还原现场情况，以检测安装节点可行性及难度。

HANNAH
House of Cores
Houston, USA 2022

HANNAH
核心之屋
美国，休斯顿 2022

The House of Cores is the first multi-story 3D-printed building in the United States. This 375-square-meter home, featuring a 12-meter tall chimney, showcases the rapidly expanding design and construction possibilities of 3D printing technology and mass customization in the field of architecture.

By coupling innovations in concrete 3D printing with wood framing techniques, a structurally efficient, easily replicable, and materially responsive building system has been created. This approach allows the two material systems to be used strategically and aims to increase the applicability of 3D printing in the US, where framing is one of the most common construction techniques.

The project design began with the objective of developing a 3D printing approach that could be scalable and applicable for multifamily housing in the future. The building is conceptualized as a series of 3D-printed load-bearing cores that contain functional spaces, bathrooms, storage and stairs.

The spatial cores are connected by wood framing to produce an architectural alternation of concrete and framed interiors. Every aspect of the building, from the overall spatial configuration to the scale of architectural detailing such as shelving and openings, was informed by the unique fabrication logic of 3D printed concrete.

Using the modular COBOD BOD2 gantry printer, the project takes advantage of the printer's modularity for its design layout. The project's scalable design and construction process is aimed to be translatable and applicable for multifamily housing and mixed-use construction.

Credits and Data
Architectural design and project planning: HANNAH
3D construction printing solutions: PERI 3D Construction
Engineering & general contracting service: CIVE.inc
Industry Partners: Quikrete, Huntsman Building Solutions, Carrier, Simpson Strong-Tie, Kohler

pp.120-121: House of Cores interior elevation.
This page, above: Aerial view of printing in progress. This page, below left: Print path of customized shelving. This page, below right: Close up of printed wall. Opposite: Visualization of the north façade.

第 120-121 页：室内立面。
本页，上：鸟瞰打印的过程。本页，左下：定制橱柜的打印路径。本页，右下：打印墙体的近景。对页：北立面的效果图。

This page: View of 2-story printed structure. Opposite: View through wall openings on the ground floor.
本页：两层打印的结构。对页：从一层望穿门洞。

核心之屋是美国第一座多层 3D 打印建筑。这座拥有 12 米高烟囱的 375 平方米住宅，展示了 3D 打印技术和大规模定制在建筑领域迅速扩展的设计和建造可能性。

通过将混凝土 3D 打印的创新与木框架技术相结合，创建一个结构高效、易于复制和材料响应的建筑系统，这种方法允许战略性地使用这两种材料系统，旨在提高 3D 打印在美国建筑领域的适用性，框架是最常见的建筑技术之一。

该项目设计的目标是开发一种可扩展且适用于未来多户住宅的 3D 打印方法。该建筑被概念化为一系列包含功能空间和楼梯的印刷要点。空间核心由木框架连接，以产生混凝土和框架内部的建筑交替。建筑的每个方面，从整体空间配置到建筑细节的尺度——例如架子和开口——都采用了 3D 打印混凝土的独特制造逻辑。

该项目使用模块化 COBOD BOD2 龙门打印机，利用打印机的模块化设计布局。该项目的可扩展设计和施工过程旨在转化并适用于多户住宅和混合用途建筑。

HANNAH
Ashen Cabin
Ithaca, USA 2019

HANNAH
梣木小屋
美国，伊萨卡 2019

Ashen Cabin is an innovative building that is 3D printed from concrete and clothed in a robotically fabricated envelope made of Emerald Ash Borer (EAB)-infested timber. This project aims to address the environmental problem caused by this beetle in North American forest ecosystems. Due to the challenging geometries of the affected Ash trees, they cannot be processed by regular sawmills and are treated as waste wood or used as low-value "firewood". Ashen Cabin upcycled the infested waste wood into a sustainable and affordable building material using high-precision 3D scanning and robotic fabrication technology.

Digital design and fabrication technologies play a pivotal role in the creation of Ashen Cabin, enabling new material methods, forms of construction, and architectural design languages. The main concrete structure is raised on 3D printed concrete legs and was fabricated using a custom 3D printing process. It consists of three programmatic areas - a table, a storage seat element, and a 20' tall working fireplace - and is characterized by expressive and functional corbelling motifs.

The irregular tree logs used for the building's facade are 3D scanned and sawn into naturally curved boards of various thicknesses using an industrial robotic arm with a custom band saw end effector. These boards are then arrayed into interlocking SIP facade panels and insulated with a two-component closed-cell foam. The resulting facade assembly is fully ventilated, detailed to manage shrinkage, and does not require an additional rain screen.

Ashen Cabin strikes a balance between familiarity and novelty, combining advanced technology with elemental design features. The curvature of the wood surfaces is strategically deployed to highlight architectural elements such as windows, entrance canopy, and roof drainage, while also offering additional programmatic opportunities such as integrated shelving, desk space, or storage. Despite being transformed and reconfigured, the natural tree remains legible in the overall architectural design.

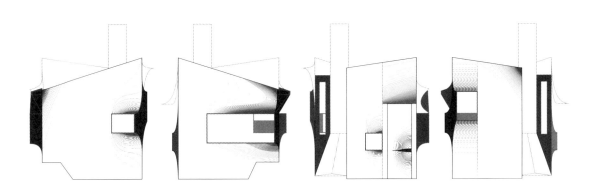

Elevation diagrams highlighting the surface curvature / 显示表面曲率的立面图

pp.126-127: Northeast corner of cabin.
This page: Southeast view of cabin. Opposite: Northwest view of cabin and 3D printed chimney within the landscape.

第 126-127 页：小屋的东北角。
本页：小屋的东南角。对页：小屋的西北角与 3D 打印的烟囱。

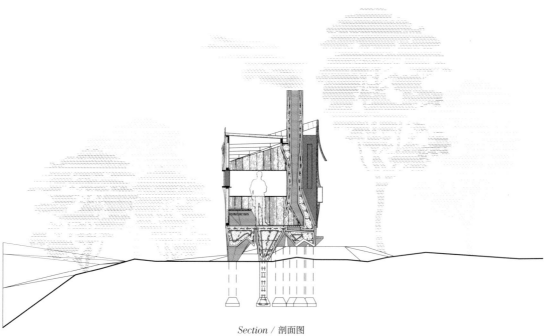

Section / 剖面图

This page, Opposite: Interior views.

本页，对页：小屋室内。

"梣木小屋"是一座使用混凝土3D打印并包裹着被翡翠灰螟（EAB）侵害的木材制成的外壳的创新建筑。该项目为解决北美森林生态系统中翡翠灰螟甲虫造成的大规模环境问题开辟了一条新途径。在多达数十亿棵受影响的白蜡树中，有大量的要么作为排放二氧化碳的"有机废物"留在森林中腐烂，要么被用作低价值的"木柴"，因为它们复杂的几何形状，无法由常规锯木厂加工。通过实施高精度3D扫描和机器人制造技术，将被翡翠灰螟侵害的"废木材"升级改造为一种供应充足、价格合理且可持续的建筑材料。

数字设计和制造技术在"梣木小屋"制造中起了关键作用，它们促进了新的材料运用方法、构造结合方式、建构形式和建筑设计语言。主要的混凝土结构是通过定制的3D打印工艺制造的，并立在3D打印的混凝土底座上。混凝土结构结合了三个特别设计的部分，一张桌子、一组置物椅和一个20英尺（约6.096米）高的工作壁炉。梁托结构成为一个富有表现力和功能性的主题，以突出3D打印混凝土的层叠特性。

可以对不规则的原木进行3D扫描，并使用带有定制带锯附件的工业机械臂，将其锯成各种厚度的自然弯曲板。这些板排列成互锁的结构承重保温板（SIP）立面面板，并使用双组分闭孔泡沫绝缘。由此产生的立面组件完全通风，经过细节处理以应对收缩，并且不需要额外的雨幕。

在建筑上，"梣木小屋"走在熟悉和陌生之间；介于先进技术和基本设计特征之间。木质外壳的曲率被战略性地部署以突出建筑重要部件，例如窗户、入口处的雨篷、屋顶排水系统，或提供额外的使用功能，例如集成的搁架、书桌或存储空间。在改造和重新配置的同时，天然树木在建筑设计中仍然清晰可辨。

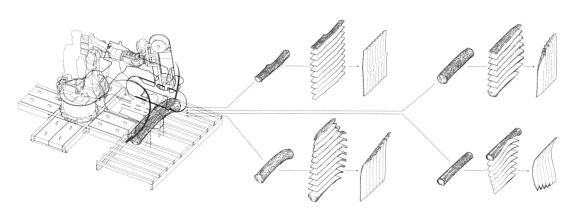

Fabrication diagram: robotic slicing of irregular wood geometries, tool paths, and corresponding surface conditions.
制造图示：不规则木材几何形状的机器切割、工具路径和相应的表面状况。

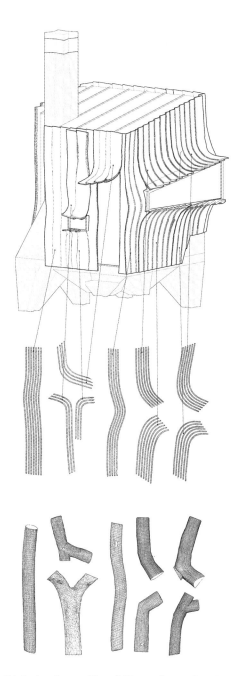

Fabrication diagram: View of chimney, door, and awning.

制造图示：烟囱、门和雨篷。

This page, above: Architectural articulations and corresponding log geometries. This page, middle: Printing of concrete leg modules with reusable gravel support. This page, below: Robotically cut façade prototype. Opposite: View of chimney, door, and awning.

本页，上：建筑表达与相应的原木几何形状。本页，中：使用可重复使用的砾石支撑打印混凝底部模块。本页，下：机器切割的立面原型。对页：烟囱、门和雨篷的视图。

Onesight Technology
Qingdao Ruyi Lake Complex
Qingdao, China 2022

以见科技
青岛如意湖
中国，青岛 2022

This project is located in Qingdao, Shandong Province, with a total building area of 170,800 square meters, consisting of 89,100 square meters above ground and 81,700 square meters in the basement. The building height is 39 meters for the complex and 18.95 meters for the element exhibition area. Despite the tight construction schedule of 235 days, the use of BIM+AR technology through the Onesight BIM+AR Construction Assistant helped ensure that the project was completed on time and with high quality.

During the project completion simulation, site embedded pipeline disclosure, mechanical and electrical pipeline acceptance, reserved hole recheck and acceptance, steel structure recheck and acceptance, mechanical and electrical equipment installation disclosure, and curtain wall completion simulation, the Onesight BIM+AR Construction Assistant was utilized. The use of AR technology overlapped BIM model data and the real world, breaking down communication barriers between the model and the Ruyi Lake construction site. This allowed for the building of a digital bridge for two-way information feedback between the BIM model and the construction site, thereby maximizing the value of BIM data in the entire building life cycle.

项目位于山东省青岛市，建筑面积17.08万 m²，地上8.91万 m²，地下室 8.17 万 m²，综合馆建筑高度39m、元素展示区建筑高度 18.95m。项目总工期 235 天，在工期非常紧张的情况下，BIM+AR 技术助力此项目按时、高质量的完成。

一见 ®AR·施工助手被运用于此项目建成模拟、场地预埋管线交底、机电管线验收、预留洞复核验收、钢结构复核验收、机电设备安装交底、幕墙建成模拟的过程中，提升了施工质量和效率，减少返工损失，保证项目在工期内完成。利用 AR 技术，使 BIM 模型数据与真实世界深入交叠，打破模型信息传递的桎梏，解决了模型与如意湖施工现场沟通不畅的问题，为 BIM 模型和施工现场架起了信息双向反馈的数字化桥梁，发挥了 BIM 数据在建筑生命全周期的价值。

1. Project Simulation: By applying the Onesight BIM+AR Construction Assistant, BIM+AR technology, and scanning the model location QR code, the project's BIM model can be projected 1:1 onto the real scene. This allowed for the simulation of the project's completion on-site, providing a reference for the owner and designer in the project planning and scheme comparison stage. Additionally, it can also simulate the progress with the construction schedule, providing a basis for construction management personnel in the field layout and progress control.

1. 建成模拟：使用施工助手软件，应用 BIM+AR 技术，扫描模型定位二维码，即可将项目外观 BIM 模型 1:1 的投射到现实场景。利用初期 BIM 模型，在项目实地进行建成模拟，可以为业主和设计方在项目规划和方案比选阶段提供参考，同时还可以配合施工进度计划进行进度模拟，为施工管理人员在现场布置、进度把控方面提供依据。

2. Visual Construction Disclosure: Before construction, visual construction disclosure can be carried out for the embedded pipelines in the outdoor site. The route and position of the pipeline can also be set out and positioned with the total station. During construction, the position of the constructed pipelines can be rechecked for quick and intuitive comparison and acceptance. The Onesight BIM+AR Construction Assistant can generate the rectification form in real-time, notify the relevant personnel, and set the rectification period to rectify traceable problems, providing visual and traceable digital data for subsequent operation and maintenance links. After construction, AR projection can be carried out on the completed model through the Onesight BIM+AR Construction Assistant, and the pipelines of concealed works will be at a glance, providing a basis for later project maintenance, reconstruction, and excavation.

2. 场地预埋管线交底：施工前，可对室外场地预埋管线进行可视化施工交底。也可以结合全站仪，对管线的走向、位置进行放线与定位。施工中，对已施工的管线进行位置复核，进行快速直观的比对验收，还可以在施工助手软件中实时生成整改单，通知相关人员并设定整改期限，进行可追溯的问题整改，为后续运维环节提供可视化、可追溯的数字资料。施工后，可通过施工助手对竣工模型进行 AR 投射，隐蔽工程管线将一览无余，为后期项目维护、改造、开挖提供依据。

3. Acceptance of mechanical and electrical pipelines: The Onesight BIM+AR Construction Assistant is utilized with BIM+AR technology to scan the model location QR code, projecting the electromechanical BIM model 1:1 to the actual site. The model's transparency is adjusted according to the specialty, allowing the comparison and checking of the matching degree between the model and the on-site construction. This enables quick and intuitive comparison and acceptance of the arrangement, trend, size, height, etc. of the pipeline and equipment after installation. Any inconsistencies can be detected, and the Onesight BIM+AR Construction Assistant can immediately generate a rectification form, notify relevant personnel, and set a rectification period to rectify traceable issues. With these measures, operations before construction are assured, and errors or omissions during construction are minimized, ensuring that data after construction is reliable.

3. 机电管线验收：使用施工助手软件，应用 BIM+AR 技术，扫描模型定位二维码，即可将机电 BIM 模型 1:1 地投射到现场场景。按专业调整模型透明度，比对查看模型与现场施工情况的匹配程度，在安装完成后对的管道和设备的排布、走向、尺寸、高度等进行快速直观地比对验收。如发现不一致，可在施工软件助手中实时生成整改单，通知相关人员并设定整改期限，进行可追溯的问题整改。通过这种方式，实现施工前作业无疑虑，施工中工序无错漏，施工后数据有依据。

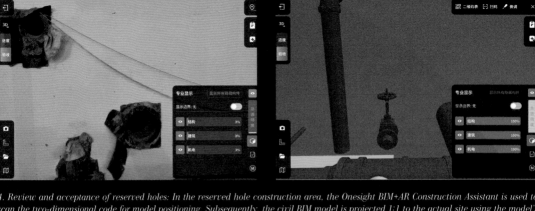

4. Review and acceptance of reserved holes: In the reserved hole construction area, the Onesight BIM+AR Construction Assistant is used to scan the two-dimensional code for model positioning. Subsequently, the civil BIM model is projected 1:1 to the actual site using the model's positioning. The civil engineering BIM model is used to compare and recheck the civil engineering reserved holes on the construction site, according to the actual construction completion. At the stage of formwork erection and reservation or after formwork removal, the quantity, position, and size of embedded sleeves and reserved holes can be checked. If any inconsistencies are found, the Onesight BIM+AR Construction Assistant can generate a rectification form in real time, notify relevant personnel, and set a rectification period to rectify traceable problems.

4. 预留洞复核验收：在预留洞施工区域，使用施工助手软件，扫描模型定位二维码，即可将土建 BIM 模型 1:1 地投射到现实场景。根据实际施工完成情况，利用土建 BIM 模型，对施工现场土建预留孔洞进行比对复核。可以在支模预留阶段或拆模后，核验预埋套管、预留洞口的数量、位置、尺寸。如发现不一致，可在施工软件助手中实时生成整改单，通知相关人员并设定整改期限，进行可追溯的问题整改。

5. Steel structure review and acceptance: To review and accept steel structure construction, use the Onesight BIM+AR Construction Assistant to scan the model location QR code, which projects the steel structure BIM model 1:1 to the actual scene. Adjust the transparency of the model for better viewing and compare it with the site to ensure the dimensions, positions, omissions, etc. of the steel structure are correct. In case of any inconsistencies, the Onesight BIM+AR Construction Assistant can generate a rectification form in real-time, notify relevant personnel, and set a rectification period to resolve any traceable issues.

5. 钢结构复核验收：在钢结构施工区域，使用施工助手软件，扫描模型定位二维码，即可将钢结构 BIM 模型 1:1 地投射到现实场景。调整专业透明度查看模型与现场的匹配程度，对钢结构安装尺寸、位置、漏项等进行快速直观地比对验收。如发现不一致，可在施工软件助手中实时生成整改单，通知相关人员并设定整改期限，进行可追溯的问题整改。

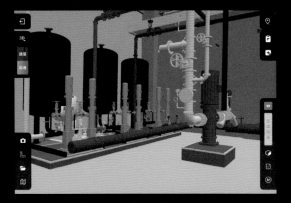 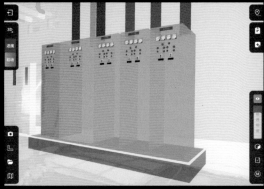

6. Installation of equipment in the computer room: Use the Onesight BIM+AR Construction Assistant, apply BIM+AR technology, and scan the model location QR code to project the electromechanical BIM model 1:1 to the real scene. Adjust the transparency of the model to compare it with the on-site construction and accept the arrangement, trend, size, height, etc., of the pipeline and equipment after installation. In case of any inconsistencies, the Onesight BIM+AR Construction Assistant can generate a rectification form in real-time, notify relevant personnel, and set a rectification period to resolve any traceable problems. This way, you can eliminate any doubts before construction, minimize errors or omissions in the process of construction, and ensure the accuracy of data after construction.

6. 机房设备安装交底：使用施工助手软件，应用 BIM+AR 技术，扫描模型定位二维码，即可将机电 BIM 模型 1:1 地投射到现实场景。按专业调整模型透明度，可分专业显示来进行可视化交底，使各班组在施工前就能明确各专业的空间位置关系，避免施工过程中各班组因"抢位置"，从而造成不必要的返工。

7. Simulation of curtain wall construction: Before installing and constructing a curtain wall, carry out a simulation of the curtain wall installation and completion in the construction area, and review the rationality and matching of the installation beforehand by using BIM+AR visualization tools. Based on the original plan, provide an intuitive effect of the curtain wall installation and construction to construction management personnel to enable them to offer constructive suggestions as early as possible and ensure a smooth construction process.

7. 幕墙建成模拟：幕墙安装施工前，对施工区域进行幕墙安装建成模拟，利用 BIM+AR 可视化手段提前复核幕墙安装合理性、匹配性，在原有方案的基础上，提早对幕墙的施工方案进行警示，呈现直观的效果，便于施工管理人员及早提出建设性的意见，为施工保驾护航。

FABO
Casablanca Biennale —"Chopsticks"
Casablanca, Morocco 2012

"数制"工坊
卡萨布兰卡艺术双年展——"筷子"
摩洛哥,卡莎布兰卡 2012

CHOPSTICKS

Chopsticks serve as essential utensils for daily use in Asian countries like China, Japan, and Thailand. The intricate arrangement of paired chopsticks and rubber bands creates repeating triangular geometric patterns, reminiscent of the organic growth of leaves in nature. This design gives rise to three-dimensional geometric patterns that extend the visual horizon when placed within natural settings, such as beaches, sandy landscapes, or lush bushes. Drawing inspiration from traditional Arabic patterns and crafted using chopsticks, this form establishes a captivating dialogue between Eastern and Western cultures, as well as between culture and nature. (Casablanca Biennial)

The "Cloud" is a complex geometric structure constructed from thousands of pairs of chopsticks, which are the daily utensils used by Asian families. Expanding through space, the Cloud represents not only a natural element but also functions as an information exchange platform that facilitates data sharing through an invisible network. "Chop-Cloud" repurposes discarded chopsticks collected by our team, mimicking the invisible cyber world that exists in the air. (Shanghai 353 Plaza)

pp. 138-139, This page: "Chopsticks" exhibited at 1st Casablanca Biennale (2012). Opposite: "Chop Cloud" exhibited at Shanghai 353 Plaza (2014).

第138-139页，本页：卡萨布兰卡艺术双年展——"筷子"（2012年）。
对页：上海353广场装置展——"筷云"（2014年）。

筷子

筷子是中国、日本和泰国等亚洲国家日常使用的基本工具。由一对筷子和橡皮筋构成的结构图案创造了重复的三角形几何图案,类似于自然界中叶子的生长方式。这种设计在自然环境中(如海滩、沙滩或灌木丛)中放置时,形成了空间中延伸视野的立体几何图案。这种形式基于传统的阿拉伯图案,由筷子构成,创造了东西方文化之间、文化与自然之间的对话。(卡萨布兰卡艺术双年展)

"云"是由成千上万对筷子构成的复杂几何形状,这些筷子是亚洲家庭日常使用的工具,它们在空间中扩展。云是一种自然元素,但也被称为一个通过无形网络共享数据的信息交流平台。"筷云"重新利用我们团队收集的废弃筷子,模仿存在于空中的无形网络世界。(上海 353 广场)

MOBILE FABLAB O (FABO)

The world is becoming increasingly digital, and digital profiles are shared globally but manufactured locally, which essentially affects local and global manufacturing landscapes.

A Fab Lab, founded by the Massachusetts Institute of Technology (MIT), is a modular smart manufacturing laboratory. It is not only a place where everyone can make their own products but also a platform to promote the new consumption mode - "self-sufficient making" that reduces energy waste caused by mass-production chains in the industrial era. Fablab O is the first built Fab Lab in mainland China.

Designed by Professor DING, Mobile FABO is an attempt to modularize our space and content into a standard container, making it moveable anywhere in the world! The scalable structure makes the space extensible, forming multiple functional layouts through different combinations. Cloud provides efficient data exchange paths between people and people, machines and machines, and people and machines. In order to build a new distributed manufacturing network, Fablab O, together with related organizations and departments as well as investors, is planning to build another 1000 Fab Labs to form the distributed manufacturing network in China.

This page, Opposite: Mobile FabLab O (FABO) & Product Instruction (2019).

本页，对页："数制"魔方实景照片及产品说明书（2019 年）。

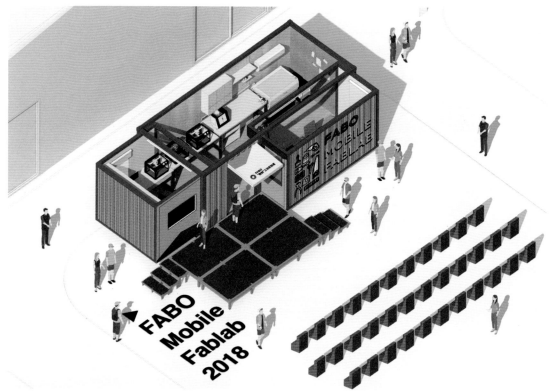

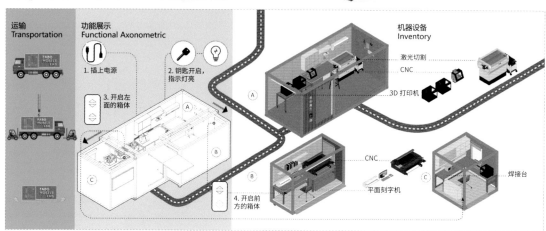

"数制"魔方

世界正在数字化。全世界数字互联，制造本土化——这将本质上影响全球和本土的格局。

Fab Lab 是创始于美国麻省理工学院（MIT）的一个模块化的智能工厂，在这里，不仅促进个人创新数制的呈现，也提出了减少工业时代标准生产流通带来能源浪费的全新消费模式——自造。Fablab O（简称 FABO）|"数制"工坊是中国大陆建立的第一个 Fab Lab 实验室。

Mobile Fablab O 是可移动、灵活变形的"数制"魔方，是 FABO 空间和内容模块化的尝试，魔方以标准集装箱模块为载体，通过运载，可以移动在全世界的任何角落；可伸缩的结构让空间具有延伸性，通过不同组合形成功能多样布局；数据云端互联，形成人与人、机器与机器、人与机器的高效数据交换途径。FABO 中国数字工坊联合相关部门和机构，在中国打造 1,000 个这样的实验室，编织出一种全新的中国分布式智造的数制网络。

STE(D)M Education Kit Package

The cultivation of innovative and scientific talent is the foundation of national competitiveness in the 21st century. China is currently undergoing a transformation and upgrade from "Made in China" to "Created in China", and the talent strategy of innovation-driven development has resulted in the rapid spread of the concept of "STEM" throughout the country. Professor Ding Junfeng's team uses design discipline in universities to promote STEM education through "Design Thinking-D". This approach guides STEM across disciplinary scenarios and project-based learning, enhances students' understanding of real problems, and strengthens the experiential learning process, thereby developing students' internal drive and innovative motivation.

STE(D)M education, led by design disciplines, is a system that includes not only curriculum content, but also innovation space, educational tools, and more. Design thinking is better equipped to take advantage of top-level design and system design for innovation education. The following two cases are experiential class curriculum kits designed by the team to match the educational goals of STE(D)M as an essential element of the science classroom after the double reduction in basic education.

BOOKi Project is a series of kit packages designed by Mr. Ding Junfeng's team for the DEMO curriculum of youth STE(D)M science and technology education. The idea for the project comes from flip-page reading experience. By opening BOOKi, the parts used in the course are displayed on a long flat scroll, making teaching and operation clear at a glance. This also reduces the risk of losing parts, while significantly increasing the fun and efficiency of the assembly process. The modular design cleverly addresses the diverse needs of the series of courses with different combinations of parts. The packaging uses recyclable kraft paper as the primary material, and the outer packaging material is also used in the project production process to reduce material waste and strengthen youth's awareness of sustainability and environmental protection.

Shanghaino is a simplified version of the popular Arduino microcontroller, developed by Fablab O Shanghai. It allows students from K12 to university to learn electronic production, PCB soldering, programming, and prototyping a circuit in a fun and engaging way. They have to assemble their own device!

It is based on the "Arduino on a Breadboard" project and other open-source projects like the Diavolino from Evil Mad Scientist and Fabschoolino from Fablab Amsterdam. All of them originated from the original Arduino project, which began in 2003 as a program for students at the Interaction Design Institute in Ivrea, Italy. The aim was to provide a low-cost and easy way for novices and professionals to create devices that interact with their environment using sensors and actuators.

Like the original Arduino, the Shanghaino is also an open-source project, and its peculiarity is the style of the PCB, cut in the shape of the city of Shanghai.

The kit has few and simple components and can be assembled and welded in less than an hour by anyone. Its specific design makes it easy to assemble and solder, even for a complete beginner. As soon as it is connected to a computer, it can be programmed in hundreds of different ways. It is fully compatible with the Arduino environment and is perfect for teaching coding, physical computing, electronics, or just for playing around!

The creative design of primary school students after assembling Crazy Mario (the brush car).

小学生组装疯狂马力欧（电刷小车）后的创意设计。

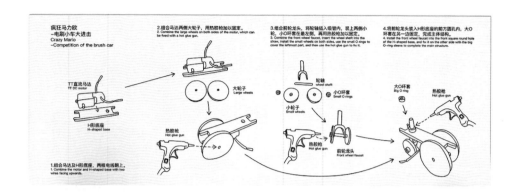

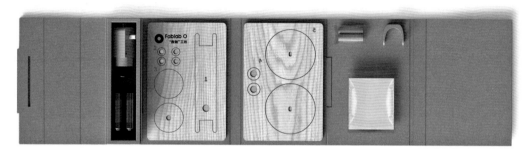

BOOKi-the packaging design of Crazy Mario (the brush car)
BOOKi——疯狂马力欧（电刷小车）外包装套件设计

STE (D) M 教育套件包

创新科技型人才的培养是 21 世纪国家竞争力的基础。我国目前正处在一个由"中国制造"向"中国创造"转型升级的阶段，创新驱动发展的人才战略让"STEM"理念得以在中国大地迅速传播。丁峻峰教授团队借助高校设计学科，以"设计思维—D"指引 STEM 跨越学科场景和项目式学习，推进 STE (D) M 教育，增进学生对于真实问题的理解并强化体验式学习过程，从而发掘学生学习的内驱力和创新能动性。

设计学科引领的 STE (D) M 教育是一个系统，不仅仅包含课程内容、创新空间、教育工具等，设计思维更能发挥创新教育顶层设计和系统设计的优势。下面的两则案例就是团队为了配合 STE (D) M 教育目标而设计的体验课课程套件，作为基础教育双减以后科普课堂的重要内容。

BOOKi 项目，是丁峻峰老师团队针对青少年 STE (D) M 科创教育 DEMO 课程设计的系列套件包。项目的创意来源于一种开本阅读的折页体验，通过打开 BOOKi，课程所用到的零部件都展示在一个平铺的长卷上，不仅让教学和操作一目了然，而且减少了遗失零件的风险，同时极大增添了组装过程的趣味和效能。模块化的设计巧妙解决系列课程的零件的不同组合下的多样化的需求。包装采用可回收牛皮纸作为主要材料，外包装材料也被运用于项目制作过程，减少了材料浪费，强化了青少年对于可持续和环境保护的意识。

当前网红的 Arduino 微控制器有上海地图版了——"小上海（Shanghaino）"，这是丁教授团队——中国"数制"工坊在上海开发的开源电路板。有了这款产品，从小学生到大学生均可以通过一种有趣的方式来学习电子制作、焊接、电路编程和原型设计，以此来智造万物，并且还能参与其中：他们可以动手组装自己的 Arduino 主板并开始编程。

"小上海"是基于"Arduino 面包板"、Diavolino 的疯狂科学实验室和阿姆斯特丹大学创新实验室 Fablab Amsterdam 开发的 Fabschoolino 的项目。所有这些项目的原型就是 Arduino 项目，该项目发起于 2003 年，当时是意大利伊夫雷亚（Ivrea）的互动设计学院为学生设计的一门课程，该课程的初衷是，让初学者和专业人员可以用一种低成本且简便的方式，来创造通过传感器和制动器与环境互动的设备。

和其原型 Arduino 一样，"小上海"是一个开源项目-印刷电路板（PCB），但其独特的外观设计（上海地图），让其成为名副其实的"上海派"。

这套设备的组件少而简单，不管是谁，只要花上一小时不到的时间就能完成。由于其特殊的设计，即使是一个完完全全的初学者也可轻松将其组装和焊接。我们只要把组装好的设备连到电脑上，就可以用上百种不同的方式对其进行编程：该设备与 Arduino 环境完全兼容，可完美地用于编程、物理运算、电子学等方面的教学，或者只是玩玩游戏也是非常不错的。

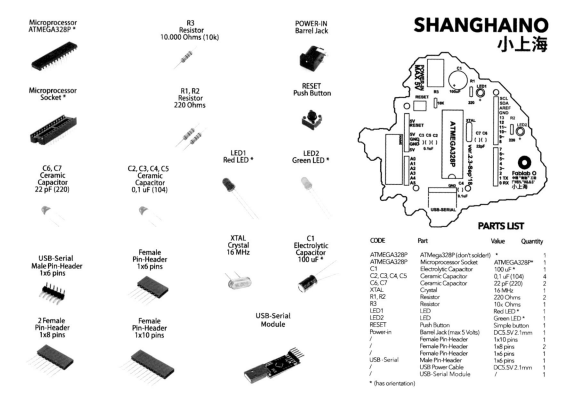

This page: Instruction of Shanghaino (2018). Opposite: Shanghaino and programmable circuit board after soldering (2018).

本页："小上海"的使用说明（2018 年）。对页："小上海"套盒及焊接好的可以编程的电路板（2018 年）。

Essay:
Software Driven Robotics for Intelligent Construction

LIANG Zee / LAI Kuan-ting / MENG Hao

论文
面向智能建造的新型机器人技术
梁喆 / 赖冠廷 / 孟浩

Current Situation in the Domestic Construction Manufacturing Industry

According to data released by the National Bureau of Statistics of China, the total output value of the national construction industry in 2021 was 29,307.9 billion yuan. However, despite being a large player in construction, China still faces many weaknesses and problems in the construction manufacturing industry. One of the main problems is the low level of informatization and automation in the industry. Information is often fragmented, and there is no platform for unified management of component production. Processing information is disassembled and distributed from construction drawings through subcontractors, and then processed by workers using terminal equipment. This can lead to many problems being accumulated, lagging or hidden until they are finally exposed on-site, resulting in rework and increased costs. The foundation of domestic construction industrialization is also weak, and most enterprises still rely on expensive imported equipment. To reduce costs, building factories often rely heavily on skilled workers' experience, leading to many omissions and processing errors. In the construction industry, labor costs are increasing year by year, and the age structure is getting older. Fewer young people choose to enter the construction and related manufacturing industries, leading to a brain drain of professional counterparts. Automation has been popularized in advanced manufacturing industries, with some factories and production lines even reaching unmanned configuration. Tesla's super factory, for example, produces one Model Y every 34 seconds thanks to reasonable planning, design, and high degree of intelligent automation. Currently, 100% automatic production of Airbus' large door frame parts has been achieved in the Varel factory in Germany, and the whole process is completed by robots without manual intervention. Airbus also has the world's most modern fuselage structure integrated production line in Hamburg, Germany. However, compared to these high-end manufacturing industries, the level of automation in the construction manufacturing industry is still in its infancy. The automated mass production of large-scale standard parts cannot meet the complex requirements of customization and flexibility in the production of building components. The diversity of materials, processes, and nodes makes it difficult to unify the production process, and external factors such as environment, site, and construction period make construction projects extremely complex. All of the above are strong resistances on the road to automation in the construction manufacturing industry.

Technological Revolution in the Construction Industry - New Industrial Software

Industrial software is the backbone of intelligent construction in advanced manufacturing industries. In the construction industry, high-quality BIM (Building Information Model) software is widely available. However, such software has always struggled to cope with complex construction and manufacturing situations. It cannot directly translate design drawings into programming

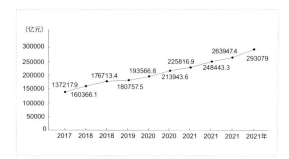

Fig.1.1 The Annual Total Value of Building Industtry in China

图 1.1 2012 年至 2021 年全国建筑业总产值

Fig.1.2 Current Condition of Building Industry

图 1.2 建造行业现状

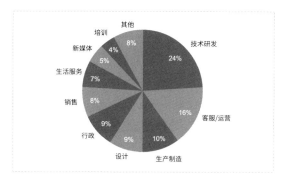

Fig.1.3 The Ratio of The Expectied Employment Position for College Graduates in 2021

图 1.3 2021 届智能制造相关专业大专毕业生期望岗位分布

Fig.1.4 AIRBUS' Fuselage-structure Assembly Line in Hamburg

图 1.4 空客在德国汉堡的机身结构生产线

国内建筑制造业现状

虽说中国是建造大国，据国家统计局公布数据[1]，2021年全国建筑业总产值293,079亿元（图1.1），但我国并不是建造强国，在建筑制造业中仍存在许多痛点和问题。如建筑制造业信息化及自动化水平低，缺乏统一管理的构件生产平台，许多问题被积累、滞后或隐藏，直至到施工现场才暴露，造成返工及成本的增加（图1.2）。其次，建筑工业化基础薄弱，主要依赖昂贵的进口设备。大部分建筑工厂以降低成本为前提，生产过程依赖熟练工人经验，常常在生产过程中出现疏漏及加工错误。再者，在建筑制造业中人力成本逐年提高，年龄结构逐年增大[2]。年轻人更倾向于进入其他行业（图1.3），专业对口人才流失相当严重[3]。在其他先进制造行业中，其实自动化早早普及，部分工厂和生产线甚至达到无人配置。特斯拉的超级工厂，平均每34秒便有一辆MODEL Y的生产完成。惊人产能的实现有赖于特斯拉工厂的高度智能化、自动化。空客（AIRBUS）的大型门框部件的100%自动化率生产已经在德国Varel工厂中实现，全程由机器人完成，无需人工干预，并在德国汉堡拥有全世界最先进的机身结构一体化生产线（图1.4）[4]。反观建筑制造业，与上述高端制造行业相比，自动化水平仍处于起步阶段。即便是自动化、批量化的大规模标准件生产方式能够达到相应水平，也无法满足建筑构件生产中定制化、柔性化的复杂要求；材料、工艺、节点的多样性的生产工艺难以统一；建筑工程项目面临环境、场所、工期等外部复杂因素。以上都是建筑制造业自动化道路上的强大阻力。

Fig.2.1 Digital Project from Gehry Technologies

图 2.1 铿利科技公司开发 BIM 软件 Digital Project

Fig.2.2 RoBIM from RoboticPlus.AI

图 2.2 大界的建筑工业软件 RoBIM

language that mechanical equipment can recognize. Therefore, the customization of building components requires more powerful industrial software to meet the complex needs in the production of building components.

In our opinion, the construction industry is in dire need of technological change. Building production and manufacturing need to quickly learn from industrialized manufacturing. By developing intelligent building industry software, connecting the data of design and construction, and introducing advanced robot equipment in factories and construction sites, production automation, informatization, and intelligence can be achieved. This man-machine interaction mode can meet the technical level of modern workers. With quality and efficiency of the project assured, the transformation of the construction from manual work to technical work will retain and attract young construction practitioners, ensuring a healthier and more stable industry development. RoboticPlus.AI's RoBIM is a construction robot industrial software platform developed to promote technological change in the construction industry. We believe that the new generation of construction industry software needs to have the following core technologies.

Auto Graphics

Automated graphics are crucial to improving the accuracy of automated production processes. Traditional processing methods for building nodes rely on human experience, which is prone to errors and can take a long time, especially for non-standard building components with complex shapes. Construction projects often rely on BIM software, such as Revit, which cannot drive the processing of factory equipment. Therefore, the key value of automated graphics in the construction manufacturing industry lies in automatically decomposing the component information in the original building model and intelligently analyzing the geometric data. Efficient construction industry software can generate forward processing data based on geometry, compare it with the reverse data collected from external information, and dynamically evaluate, correct, and optimize the final processing data of components to ensure accuracy. It can transform into a producible model suitable for industrial construction technology.

Virtual Simulation Automation

Virtual simulation is a technology that uses a digital system to imitate another real system. It simulates and plans execution and movement in the actual physical space through a simulation test conducted in a virtual environment. Advanced manufacturing uses 3D simulation technology to plan the running path and avoid equipment collision. Through virtual simulation technology, operators can simulate and preview the robot processing path in advance in the computer's 3D virtual environment. The intelligent virtual simulation technology greatly reduces the

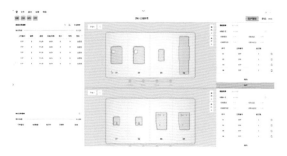

Fig.2.3 RoBIM with Auto Graphic
图 2.3 运用了自动图形学的 RoBIM 切割软件

Fig.2.4 Virtual Simulation in RoBIM
图 2.4 RoBIM 软件中的虚拟仿真交互界面

建筑行业的技术变革——新型建筑工业软件

在先进制造行业中，工业软件是串联智能建造各个环节的中枢。而在建筑行业中，即使有许多优质的 BIM 软件（图 2.1，Digital Project 是铿利科技公司 Gehry Technologies 所发展出的软件），但面对复杂的建筑制造业状况依然力不从心，因为它无法直接从设计图纸语言转化为机械设备直接能识别的程序语言。因此，建筑构件实现定制化需要更强大的工业软件来满足建筑构件生产中的复杂需求。

笔者认为建筑行业亟需一场技术变革。建筑的生产和制造需要快速向工业化制造借鉴和学习，通过开发智能建筑工业软件，联通设计端和建造端，通过数据驱动（data-driven）[5] 工厂和工地机器人装备，实现生产的自动化、信息化和智能化。同时提供人机交互模式以符合现代工人的技术水平，保证工程质量和效率，让建筑建造由体力工作转向技术工作，留住和吸引年轻有才华的建筑从业人员，使行业可以更加健康和稳定的发展。RoBIM 正是大界自创立之初就开始研发的建筑机器人工业软件平台（图 2.2），目标在于推动建筑行业的技术变革。笔者认为新一代的建筑工业软件需要具有以下几个核心技术。

核心一：自动图形学

自动图形学是提升自动化生产工艺精度的关键一环。建筑节点传统的加工方式依赖于人工经验，尤其对于形态复杂的非标建筑构件容易产生难以复核的错误，且加工时间长，效率低下。在建筑项目中，虽有 Revit 等 BIM 软件，但这些软件所提供的信息并不具备加工数据[6]，无法驱动工厂设备加工生产。自动图形学在建筑制造业中的关键价值在于自动分解原始建模中的构件信息，智能的解析几何数据。当传统三维模型中的数据进入到智能 CAM 软件的自动图形学解析过程中，程序将数据推导出加工路径，为自动化编程提供图形基础（图 2.3，RoBIM 切割软件，能够一键导入二维图纸或三维模型进行解析）。高效的建筑工业软件能够基于几何图形学的正向加工生成数据，与外部信息采集的逆向数据进行比对，从而对构件的最终加工数据进行动态的评估、修正和优化，确保其准确合理的转化为适合工业化建造技术的可生产型模型。

核心二：虚拟仿真自动化

虚拟仿真是以一个数字系统模仿另一个真实系统的技术，在虚拟环境中进行的仿真模拟测试，生产中常以三维仿真技术来计划运行路径，规避设备碰撞等。在先进制造业中，通过虚拟仿真技术，操作人员能够提前在计算机三维虚拟环境中，模拟和预览机器人加工路径。智能化的虚拟仿真技术极大减少了传统工业机器人在编程过程中的人工操作，使得机器人编程操作更简易，甚至一键即可将 BIM 模型中的构件三维数据转化为 CAM 软件中的机器人加工程序，实现加工程序编程全自动化。图 2.4 中，大界 RoBIM 软件能够在虚拟数字环境中，将路径规

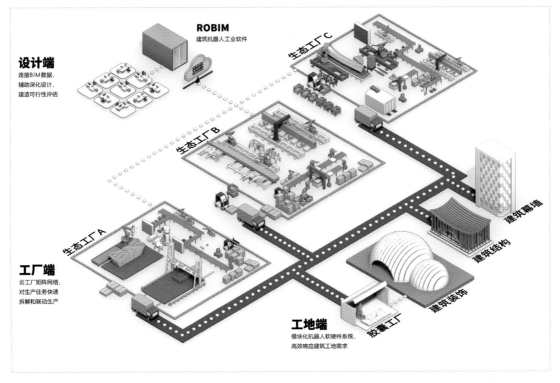

Fig.3.1 RoboticPlus.AI's Blueprint of the Intelligent Construction
图 3.1 大界智能建造未来愿景

number of manual operations in the programming process of traditional industrial robots, making robot programming easier to operate, and even converting the 3D data of components in the BIM model into CAM software with one click. The robot processing program in the machine realizes the full automation of the processing program's programming.

Digital Twin

Digital twin technology creates high-fidelity virtual models of physical objects in virtual space. Through the deployment of general-purpose robot software and hardware equipment in construction factories, digital upgrades are carried out for production enterprises, and infrastructure and digital environments are provided for production simulation. RoboticPlus.AI uses digital twin technology to simulate the production of robots in a virtual factory environment to ensure the feasibility and effectiveness of automated production.

In the smart factory project with Baoye Steel Structure, RoboticPlus.AI successfully realized true digital twin technology in the groove cutting single station. RoBIM software, combined with visual positioning, was used to automatically load and unload materials. Real-time interaction between virtual and physical can be used to predict the relative displacement deviation of virtual BIM models and actual production tools, mechanical fixtures, and actual component dimensions.

The Future of Intelligent Construction in the Construction Industry

Since the Third Industrial Revolution, the integration of intelligence and information technology has once again triggered a jump

划、智能规避和碰撞检测等进行仿真模拟，为技术人员提供设备运行预览展示，并可自动编译及导出机器人设备运行程序。

核心三：数字孪生技术

数字孪生技术是指在虚拟空间中创建物理对象的高保真虚拟模型[7]。通过在建筑工厂部署通用的机器人软硬件设备，为生产企业进行数字化升级，为生产模拟提供基础设施及数字环境。大界运用数字孪生技术在虚拟工厂的环境下进行机器人的生产模拟，以保证自动化生产的可行性和有效性。

大界在与宝冶钢构的智能工厂项目中，成功在坡口切割单站实现了真正意义上的数字孪生技术（图2.5）。大界使用 RoBIM 软件，结合视觉定位自动上下料，同时通过视觉重构，在数字三维环境中实时生成图形，拟合物理环境中的真实监测数据，运用虚拟与物理的实时交互来预测虚拟 BIM 模型和实际生产工具、机械夹具和实际构件尺寸的相对位移偏差，进行更新调整，从而自动消除了构件在工装平台上的偏离误差。在数字与物理的高速交互映射生产过程中，提供了对整个过程的实时检测分析，避免误差的累积，尤其在大尺度建筑构件的生产中克服数字模型数据或物理空间标定的不准确带来的生产精度的损失。

建筑业的智能建造未来

智能化与信息化是自第三次工业革命以来，再一次触发生产力跃升的跳板。建筑业的智能建造未来取决于能否把握这一契机。笔者认为通过为建筑生产企业提供先进的机器人自动化设备及软硬件解决方案，落地智能建造新型生产线，实现降本增效，提高自动化率。可以逐步弥补目前国内建筑构件工厂的技术短板，主动促进建筑制造业升级，从而推动建筑业整体技术转型。

同时，这种升级转型将会逐步打破建筑行业的固有格局，把建筑制造业从严重的同质化竞争转化为以技术作为衡量标准的差异化竞争。图 3.1 所示的是大界对智能建造未来愿景，大界愿意将软件系统作为平台化技术与建造企业共享，使中小企业通过共享技术有机会参与到市场竞争中，并且持续地吸纳、整合和改进行业外的新技术，倡导建筑业的协同发展。

应用一：柔性生产

在以往工厂的生产模式中，随着同质化竞争越来越激烈，能更加灵活适配市场多样化需求的制造企业将脱颖而出。聚焦到建筑制造业，在面对当今建筑构件多样化、差异化的产品需求时，传统的生产模式弊端开始凸显。为适应异形建筑构件的复杂性，建筑生产企业需要大规模生产转变为大规模定制。

对于生产企业而言，提高生产方式的技术水平实现柔性生产，以提高企业产品的灵活性和生产应变能力（图3.2）。随着技术发展，基于现今逐步成熟的自动化图形学、自动虚拟仿真等技术，建筑工厂可以更加满足建筑行业对非标定制构件及柔性化生产工艺的需求，同时也实现建筑工厂的降本增效。因此未来的建筑智能工厂，必将是以柔性生产为基础的多元定制化产品工厂。

应用二：微型胶囊工厂

除了上述提及的柔性生产外，工厂单元本身也可以突破地理空间的局限，成为更轻便、更灵活的单元。胶囊工厂（Flying Factory）是能够满足上述技术要点的便捷型移动工厂概念。不受地理位置的制约，可以快速到达工地，它由模块化建筑机器人系统及其配套硬件组成。通过新型工业软件的赋能，最大自由度地配合、最高效率地响应建设工地的要求。大界在与中建三局合作研发的项目当中，成功研发落地了国内首个以工地现场曲线钢筋弯折作为应用场景的胶囊工厂（图3.3—图3.4）。微

Fig.2.5 Virtual Simulation in RoBIM
图 2.5 大界为宝冶钢构实现真正的数字孪生

Fig.3.2 Flexible Production Line in Steel Factorie
图 3.2 某钢结构焊接工厂的柔性生产线

in productivity. RoboticPlus.AI believes that by providing advanced software and hardware solutions for robotic automation equipment to construction production enterprises and implementing intelligent construction of new production lines, costs can be reduced, efficiency increased, and the automation rate improved. This can gradually make up for the technical shortcomings of current domestic building component factories, actively promote the upgrading of the construction manufacturing industry, and thus promote the overall technological transformation of the construction industry.

At the same time, this upgrading and transformation will gradually break the inherent pattern of the construction industry, transforming it from serious homogeneous competition to differentiated competition with technology as the standard of measurement. Figure 3.1 shows RoboticPlus.AI's vision for the future of intelligent construction. RoboticPlus.AI is willing to share the software system as a platform technology with construction enterprises, so that small and medium-sized enterprises have the opportunity to participate in market competition through shared technology. It will also continue to absorb, integrate, and improve new technologies outside the industry and advocate the coordinated development of the construction industry. The goal of new intelligent construction is to help construction enterprises achieve flexible production, matrix networks, mass system management, and even flying factories.

Flexible Production

In the previous production model of factories, products were oriented and product strength was the main competitive advantage in the past. However, manufacturing enterprises that are more flexible and adaptable to the diverse needs of the market gradually stand out. Focusing on the construction manufacturing industry, in the face

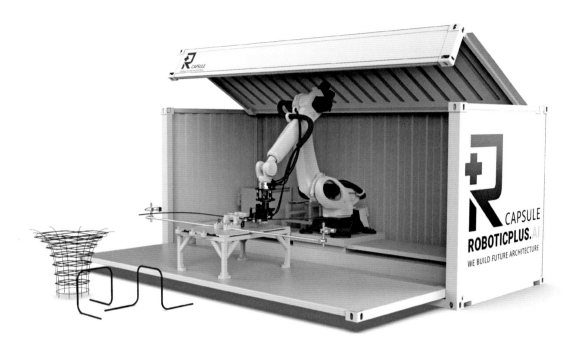

Fig.3.3 RoboticPlus.AI's Flying Factory Concept
图 3.3 大界钢筋弯折胶囊工厂概念图

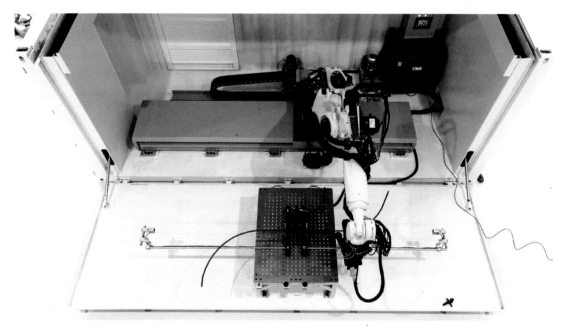

Fig.3.4 RoboticPlus.AI's Flying Factory Product On-site Manufacturing
图 3.4 大界钢筋弯折胶囊工厂产品现场

of today's diversified and differentiated product demands for building components, the drawbacks of traditional mass production have begun to emerge. In order to adapt to the complexity of specially shaped building components, construction production enterprises need to transform from mass production to mass customization.

For production enterprises, improving the technical level of production methods to achieve flexible production aims to improve the flexibility of enterprise products and production adaptability. With the development of technology, construction factories can meet the needs of the construction industry for non-standard customized components and flexible production processes, while achieving cost reduction and efficiency increases. Therefore, the future intelligent factory will be a multi-customized product factory based on flexible production.

Microcapsule Factory

In addition to the flexible production factories mentioned above, the factory unit itself can also overcome field and geographical limitations to become a lighter and more flexible unit. The Flying Factory is a convenient mobile factory concept that can fulfill the technical requirements mentioned above. It is unrestricted by geographical location and consists of a modular construction robot system and its supporting hardware. This allows it to quickly reach the construction site and its surroundings. With the help of new industrial software, the production data can be automatically decomposed to match the corresponding processing technology, overcoming factors such as materials, logistics, and labor time. This ensures the highest efficiency of the construction site and the maximum degree of freedom of cooperation.

In a cooperative R&D project with China Construction Third Engineering Bureau, RoboticPlus.AI successfully developed and produced the first microcapsule factory in China. The microcapsule factory will definitely become an indispensable part of the future of intelligent construction. Compared to traditional building production facilities, its advantages lie in its cost control, high flexibility, ease of operation, and stable production capacity of 24-hour uninterrupted production.

BIM Management System Model Combined with MES

The MES management system (Manufacturing Execution System) was proposed by the American company AMR (Advanced Manufacturing Research, Inc.) in the 1990s and is currently widely used in industrial production. The core of MES lies in the unified coordination and management of all resources and production elements to improve production operation efficiency and reduce redundant resource consumption. This plays a guiding role in the entire closed-loop production.

Currently, most of the BIM models used in the construction industry are not connected with the information data of the MES production and manufacturing system. After receiving orders for construction drawings, many construction manufacturing enterprises need to have the technical department carry out technical drawings. This process carries the risk of human error and can cause significant delays in the production cycle, especially when faced with demands for building components with a short construction period, various types, and a large number of building components. Therefore, we believe that the BIM model needs to be seamlessly connected to the MES production and manufacturing system of the building production enterprise in the future. Based on the huge database of BIM itself, combined with technologies such as automated virtual simulation and digital twinning, it needs to be converted into materials, including materials, CAM processing data, and processing information. This will support the MES platform of the construction production enterprise, carry out the production management of the whole life cycle, and finally realize the improvement of the overall energy efficiency of the construction production enterprise, achieving the effect of double improvement in quality.

Matrix Network

The future of intelligent construction in the construction industry is closely tied to the widespread adoption of matrix network production

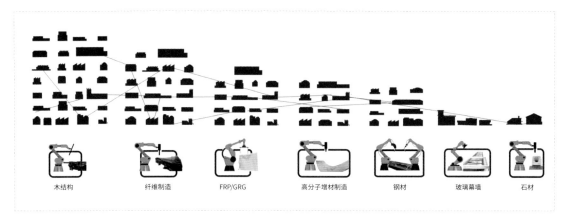

Fig.3.5 Matrix Production in Construction Manufacturing Industry
图 3.5 建筑建造业的矩阵化网络

型胶囊工厂在智能建造的未来必定是不可或缺的一部分，相比于传统的建筑生产工厂，其优势在于成本可控、灵活性强、操作简便及 24 小时不间断生产的稳定产能。

应用三：结合 BIM 模型的 MES 管理系统
MES 管理系统是由美国 AMR 公司在 20 世纪 90 年代提出，目前广泛适用于工业生产领域的生产制造执行系统。MES 的核心在于对工厂生产执行层面的协调管理，对整个工厂生产制造过程进行系统性优化，并对整个生产闭环起到指导作用[8]。

目前在建筑行业使用的 BIM 模型有大部分并未与 MES 生产制造系统进行对接，许多建筑制造企业在接收建筑施工图纸后，需要由技术部门进行二次深化，再到生产部门进行生产加工。整个过程存在信息的再转译，人为错误风险，以及会对生产周期造成拖延，尤其是面对工期短、种类杂、数量多的建筑构件需求时，对于建筑生产企业的技术部门往往是相当大的考验。笔者认为，BIM 模型在未来需无缝衔接 MES 生产制造系统，以 BIM 自身的庞大数据基础，结合自动化虚拟仿真、数字孪生的技术，转换成包含材料、工艺、加工信息的 CAM 加工数据，支撑建筑生产企业的 MES 平台，进行全生命周期的生产管理，最终实现建筑生产企业总体能效的提高，达到量质双升的效果。

应用四：矩阵化网络
建筑业智能建造的未来离不开矩阵化网络生产系统的普及。在传统建筑工厂的生产线上，每个单站只能完成一个加工任务，实现单一环节，完成标准产品的生产。而在矩阵式网络生产当中，每个站点可以搭载不同工艺的末端执行器，甚至完成产品成套加工生产，同时还能结合人工智能和机器学习等技术进行综合分析模拟，评估得到最优的生产组合措施和排产计划[9]。只有智能化的 CAM 软件、成熟的建筑 MES 和建筑 PDM 相结合才能在建筑制造业实现并推广真正意义上的矩阵化生产网络（图 3.5）。

全球智能建造实践

纵观近年的全球智能建造发展历程，国外已有多家企业、机构、实验室在智能建造的各个细分领域上实现大步迈进。美国的 Katerra 已有了制造建筑构件和产品的大型工厂，并于 2019 年将三条机器人木结构装配式生产线投入工厂使用[10]。欧洲的 Odico Formwork Robotics 利用机器人热丝切割发泡聚苯乙烯泡沫（EPS）作为混凝土浇筑模板的技术，成功克服了曲面混凝土浇筑的模具精度难题，使低成本的曲面混凝土浇筑技术能够进入建筑制造业市场。其提供的解决方案赋能了多家建筑事务所，如 GXN、Zaha Hadid Architecture、Foster & Partners。

systems. In traditional construction factories, each individual station can only perform a single processing task, limiting efficiency and productivity. However, in matrix network production, each station can be equipped with end effectors for different processes, enabling the production of complete sets of products. Moreover, the integration of artificial intelligence, machine learning, and other advanced technologies allows for comprehensive analysis and simulation to evaluate optimal production combination measures and production scheduling plans. Only the combination of intelligent CAM software, mature construction MES (manufacturing execution systems), and construction PDM (product data management) can realize and promote the full potential of matrix production network in the construction manufacturing industry.

Global Intelligent Construction Practice

Over the past few years, many enterprises, institutions, and laboratories have made significant strides in various sub-fields of intelligent construction. For instance, Katerra built large-scale factories for manufacturing building components and products, while Odico Formwork Robotics in Europe uses robotic hot wire cutting of expanded polystyrene foam (EPS) to create concrete pouring formwork. CadMakers, a construction and manufacturing technology company located in Canada, provides digital twin technology for building production and construction. Additionally, RoboticPlus AI has been committed to the development of intelligent industrial robot systems for construction robots, which has empowered the intelligent building manufacturing industry.

Conclusion

The current domestic construction manufacturing industry is facing a multitude of challenges, such as rising labor costs, aging, and a shortage of skilled professionals. The first step that construction production enterprises need to take is to learn from advanced industries and manufacturing practices. Intelligent and information technology are the primary challenges and opportunities for most construction factories. As an advocate and practitioner of intelligent construction reform, RoboticPlus.AI will continue to absorb the backbone of the industry both domestically and internationally, strengthen the research and development team, and invest in the forefront of the upgrading of the building intelligent industry.

The author believes that in the near future, RoboticPlus will release a new generation of construction industry software. This software will empower the upstream and downstream of the construction industry by providing mature software and hardware solutions for construction production enterprises and improving the automation rate and production energy efficiency. Through graphics automation, the digital twin, and other technologies, a platform for flexible production and a matrixed network can be built to enable the gradual transformation of the construction manufacturing industry from labor-intensive to industrialized intelligent production. The mature construction industry software will connect the front-end design and planning of construction projects vertically and the back-end construction management horizontally, enabling the intelligent industry upgrading and empowerment of the whole construction industry. The construction industry will truly enter the era of Industry 4.0!

加拿大的建筑和制造技术企业 CadMakers 为业主、开发商建筑施工企业提供建筑生产及施工的数字孪生技术，其 SaaS 产品对北美市场产生持续深远的影响。大界自 2016 年成立至今，也一直致力于开发建筑领域的智能化工业机器人系统，通过深耕建筑机器人的控制软件、智能算法与人机交互等核心技术，为建筑智能制造业赋能。

总结

就目前的国内建筑制造业而言，不断地面临劳动力成本提升、老龄化及智能建造专业人才外流等问题。如何向先进工业、制造业学习是建筑生产企业需要迈出的第一步，而智能化、信息化是大部分建筑工厂面临的首要挑战，也是机遇。大界作为智能建造改革的倡导者和实践者，将持续吸纳海内外行业骨干，壮大硕博士研发团队，不遗余力地投入建筑智能产业升级的一级战线当中。

笔者相信，在不远的将来，大界所研发的新一代建筑工业软件能够不断地赋能建筑行业的上下游，特别是为建筑生产企业提供成熟软硬件解决方案，提高自动化率和生产能效。通过图形学自动化、数字孪生等技术，搭建柔性生产、矩阵化网络的平台，实现建筑制造业逐步从劳动密集型到工业化智能生产的转型。成熟的建筑工业软件将横向打通建筑工厂生产，纵向对接建筑项目的前端设计策划及后端施工管理，实现建筑全行业的智能产业升级与赋能，建筑行业将真正迈入工业 4.0 的时代！

参考文献：

[1] 杨曦.我国建筑业总产值持续增长 2021 年 11 省份建筑业总产值超万亿元.(2022 年 1 月 24 日). http://finance.people.com.cn/n1/2022/0124/c1004-32338530.html.

[2] 统计局网站.2021 年农民工监测调查报告.国家统计局.(2022). http://www.gov.cn/xinwen/2022-04-29/content_5688043.htm.

[3] 直聘研究院.2021 年应届生就业趋势报告.山东：BOSS 直聘,2022.

[4] DIGITAL AND AUTOMATED PRODUCTION. Retrieved from BDLI.(2020, 2). https://www.bdli.de/en/innovation_of_the_week/digital-and-automated-production.

[5] Silvia Meschini, Kepa Iturralde, Thomas Linner, er al. Novel applications offered by Integration of Robotic Tools in BIM-based Design Workflow for Automation in Construction Processes[Z]. German,2016.

[6] 前瞻产业研究院.2020 年中国建筑工程行业市场现状及发展前景分析 [DB/OL].深圳：前瞻产业研究院,2020.

[7] 张新生.基于数字孪生的车间管控系统的设计与实现 [D].郑州：郑州大学,2018.

[8] 盛步云、苏佳奇、卢其兵.面向 MES 的生产线数据采集系统的研究 [J].计算机测量与控制,2015,23(9): 3162-3164.

[9] 丁烈云.数字建造导论 [M].北京：中国建筑工业出版社,2019.

[10] McKinsey & Company.Imagining construction's digital future [EB/OL].Singapore: McKinsey Productivity Sciences Center, 2019.

LIANG Zee　Partner/Director of Architecture BU RoboticPlus.AI
LAI Kuan-ting　Partner/CEO of Architecture BU RoboticPlus.AI
MENG Hao　Founder/CEO of RoboticPlus.AI

梁　喆　大界机器人合伙人、建筑事业部总监
赖冠廷　大界机器人合伙人、建筑事业部 CEO
孟　浩　大界机器人创始合伙人

Essay:
Architect and Architecting: The Third Wave of Architecture in the Age of New Technology
FAN Ling / CHEN Xueer

论文:
建筑师和架构:新技术可能下的第三浪建筑学
范凌 / 陈雪儿

Primer

Architecture is a discipline that is continually evolving. Contemporary Chinese architecture is characterized by what is known as "triple wave superimposition," where three development trends coexist and overlap in the present time and space. The concept of "triple waves" originates from futurist Toffler, while "triple wave superimposition" was first used by Professor ZENG Ming, Chief Strategy Officer of Alibaba, to describe the features of China's current business development. This paper employs the "triple wave superimposition" framework to examine the state of Chinese architecture within the current political and economic context. The first wave is the modernist classics, which return to the essence of architectural construction; the second wave is Spectacle Architecture, which maximizes the use of media and technology; and the third wave is System Architecture, which applies network, data, and computing under the framework of Architecting. These three waves are not superior or inferior to each other but rather sequential and superimposed upon each other in a contemporary context. While the first and second waves have been analyzed and studied extensively, the third wave remains an area of ongoing academic discourse. The aim of this paper is to provide inspiration and generate discussion surrounding the third wave.

Defining Third Wave Architecture

The origins of third-wave architecture can be traced back to two different attitudes toward human-machine interaction in the mid-20th century. One view is represented by Peter Eisenman, the originator of postmodern architecture. Eisenman argued that the relationship between idea and form should be established by human (the architect), controlling the tool (the computer). According to Eisenman, it is the subjectivity of the idea that uses the tool to produce the form, rather than the tool determining the form. Eisenman believed that the advancement of machines and technology should not replace the role of man as a subject, in order to ensure that the tradition and disciplinarity of architecture not only will not vanish, but should be further defended. Another representative of the two attitudes toward human-machine interaction is Nicholas Negroponte. Negroponte, who was trained in architecture and founded the groundbreaking Media Lab at the Massachusetts Institute of Technology (MIT) in 1985, emphasized that the intelligence and plasticity of the form itself shape an information-based system that derives from the gradual concession of the conscious subject's impact (i.e., the architect) to the system's

> The reader will recognize in the following chapters an underlying theme that is anti-architect. This must not be confused with an anti-architecture bias. Each chapter removes the architect and his design function more and more from the design process; the limit of this progression is giving the physical environment the ability to design itself, to be knowledgeable, and to have an autogenic existence.
>
> — Nicholas Negroponte
>
> But I didn't know I was thinking like a computer!
>
> — Peter Eisenman

> 读者将逐渐的意识到"反建筑师"的主题，这并不应该和"反建筑学"的偏见相混淆。每个章节都会越来越多的将建筑师和他的设计功能从设计过程中移除，这个过程的边界是让物理环境可以有自己设计的能力，从而变得智慧有知和自发的存在。
>
> ——尼古拉斯·尼葛洛庞帝

> 但我当时不知道我像计算机一样思考。
>
> ——彼得·艾森曼

引子

建筑学是持续被更新定义的学科，当代中国建筑学具有"三浪叠加"的特质，即三个发展趋势同时在当下的时空共存叠加。"三浪"的观念出自未来学家托夫勒[1]，"三浪叠加"则由阿里巴巴首席战略官曾鸣率先使用，用来描述中国当下商业的特征[2]。这里借用"三浪叠加"来分析当下政治经济语境下的中国建筑学状态：第一浪是回归建筑学建造本质的现代主义经典；第二浪是充分利用媒体和技术的景观主义建筑学；第三浪则是在建筑学思维下运用网络、数据和运算的系统架构[3]。这三浪并不存在优劣，而仅为先后顺序，在当代切片下叠加在一起。笔者认同第一浪、第二浪的存在意义和价值，相关分析和研究也颇多，本文重点分析第三浪，本文目的不是建立新的建筑学体系，而只是抛砖引玉，期待产生争鸣。

第三浪建筑学的萌芽

第三浪建筑学可以追溯到两种人与机器交互的不同态度：一个观点的代表人物是艾森曼，认为观念和形式的关系应该是由人（建筑师）控制工具（计算机）建立，主体观念通过工具产生了形式，而不是由工具决定形式。艾氏认为机器和技术的进步不应该取代作为主体的人的角色，建筑学的传统和学科性不仅不会消失，而应该被进一步捍卫。另一个观点的代表人物是尼葛洛庞帝，他接受建筑学训练，早年将建筑学作为机器进行研究，并于1985年在麻省理工学院开创性地创立了媒体实验室。尼氏强调形式本身的智慧和可塑性形成了一种基于信息的系统，来源于有意识的主体（即建筑师）逐渐让位于系统与信息的互动。尼氏提出"反建筑师"的言论，模糊了建筑学和其他学科的学科边界，形成"聚合"。[4]

尼氏的媒体观念是正在兴起的第三浪建筑学的一个引子。同期还有富勒和亚历山大的颠覆性思考，让建筑学与科技、设计、创业相结合的社会氛围之间建立关联。这条线索在20世纪60年代到70年代异常活跃，随后在欧美以符号学为代表的哲学人文思维和认知论（或者可以简称为"抵抗的建筑学"，艾森曼是其代表人物）的

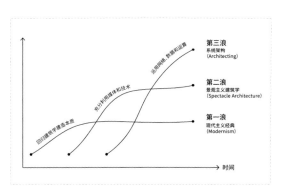

Triple waves architecture

三浪叠加——当下政治经济语境下的中国建筑学状态

interaction with information. This process reflects Negroponte's statement of "anti-architect," in which architecture, along with other disciplines, gradually blurs disciplinary boundaries to form "Convergence."

Negroponte's notion of media was an inducement to the emerging third wave of architecture. In addition to Negroponte's conceptions, the mid-20th century saw the subversive thinking of Buckminster Fuller and Christopher Alexander, who made connections between architecture and the contemporary social context that combined technology, design, and entrepreneurship. This line of thinking was particularly active in the 1960s and 1970s and was then suppressed by the development of philosophical humanistic thinking and cognitive theory represented by semiotics in Europe and the United States (or simply "the architecture of resistance," as represented by Eisenman). However, the third wave of architecture continues in cybernetics, computer science, artificial intelligence, and the Internet.

Unlike the critical nature of the first wave of architecture and the technical expressionism of the second wave, the third wave represents an architectural exploration of systematic problem solving, using System, Method, Pattern, Interaction, Behavior, and Object as the language and sustainable social (and commercial) value as the driving force.

Architecture is rapidly growing and transforming in China. China's urbanization represented by real estate development in the past 30 years has spawned both the first and second waves of architecture. In the past 10 years, urbanization has shifted from gross incremental development to stock refinement development. Along with the booming development of the Internet, social media, artificial intelligence, and big data, the development of architecture has become more open and diversified, giving birth to the budding of the third wave of architecture.

Case 1: City Brain, The Ultimate Resource Optimization Technology

One of the biggest issues facing big cities is traffic. Architects have employed various urban forms, such as overpasses, expressways, and metros, to solve this problem. City policymakers have also implemented various policies, including single and double numbers, license auctions, and more. However, these efforts have not significantly improved traffic situations. The question is, do we know exactly how many vehicles are there in the city during the daily rush hour? Is it 10%, 50%, or 100% more? Can we accurately calculate the number of vehicles moving in the city? Academician Wang Jian has raised this question and proposes using machine intelligence to solve the problem of urban design.

Urban planning and design involve the arrangement of urban resources. By counting the number of vehicles running in the city through cameras monitored at each traffic light, it was discovered that the number of cars in Hangzhou during peak hours is roughly 10% more than usual. This number can guide more intelligent and efficient regulation of traffic lights, reducing traffic congestion. This requires a computing system that can calculate the city traffic in real-time and give feedback to traffic lights for adjustment. Academician Wang Jian called this system the "City Brain," which is not the human brain of the city. Even a human brain cannot solve urban problems. The City Brain is the brain of the complex system of the city, not a single point of optimization, but the game of calculation and overall optimization of this complex system.

Using cloud computing and machine intelligence to intervene in city operations may seem like an expression of an "Authoritarian City" in the digital era. However, this problem is not only the government's responsibility as private technology companies also collect data. Recently, China's largest shared transportation company, Didi taxi, was fined more than RMB 8 billion for collecting substantial amounts of unnecessary personal data. Individual users upload their personal information entirely to various commercial applications in exchange for ease of life and commercial rewards. Governments play an integral role in regulating and legislating data

发展中偃旗息鼓。但第三浪建筑学在控制论、计算机科学、人工智能等方面的线索随后被延续到互联网的发展上。与第一浪建筑学的批判性、第二浪建筑学的技术表现主义相比，这是一条以系统性解决问题为目的的建筑学探索，以系统、方法、模式、迭代作为设计语言，并以可持续的社会价值（包括商业价值）作为进化动力。

房地产开发驱动的城市化孕育了中国第一浪和第二浪的建筑学。伴随着城市的发展从粗放增量型发展转向存量精细型发展，基于互联网、人工智能和大数据的数字经济、数字社会为第三浪建筑学的萌芽提供了土壤。这是一个进行时，我们将精读三个具有"第三浪建筑学"特征的案例。

案例一：城市大脑的终极空间资源优化

城市规划和设计是对城市资源的调配，其中交通是最头疼的问题之一。建筑师采用了各种城市形式来解决（高架、快速路、地铁等）；政策制定者则发明了各种政策（单双号、牌照拍卖等），但问题并没有得到改善。"我们是否搞清楚高峰期城市里到底多了多少车子？是10%、50%还是100%？我们有没有能力计算清楚路上有多少正在行驶的车子？"这是阿里云创始人王坚院士针对交通阻塞提出的问题。在杭州，通过每个交通灯上监控的摄像头，对城市正在运行车辆数量进行计算发现，高峰期车辆数量大概增加10%。这个数字可以指导智能摄像头更高效、实时的调控摄像头，可以把时间引入空间资源调配的维度，减轻空间上的交通阻塞。这背后需要有一套基于云计算的城市运算系统，对城市交通进行实时分析，并反馈给交通灯进行调整。王坚院士称为"城市大脑（City Brain）"，他指出：城市大脑不是城市有人的大脑，即使人的大脑也解决不了城市问题。城市大脑是城市复杂系统的大脑，不是单点优化，而是对这个复杂系统的博弈计算和整体优化。[5]

也许有人会质疑云计算介入城市管理是"集权"的表现，每个居民都被摄像头所监视，为了整体效率而牺牲个体隐私。这个问题并不能被过度简化，采集数据的不仅是政府，还有大量的私营技术企业，用户把个人信息悉数上传给各个商业应用，换取生活的便利性和商业

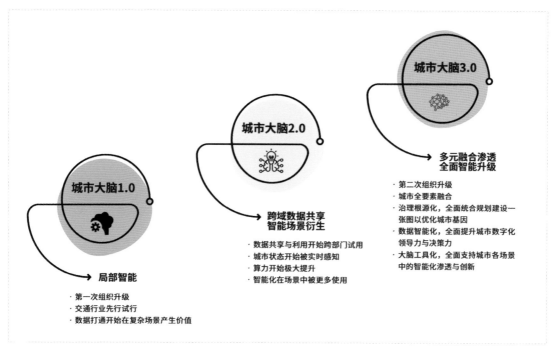

Triple waves architecture
城市大脑三次进阶过程与特征

security and AI ethics. New technological tools, such as "federal learning" technologies, may enable the exchange of data values that do not contain confidential information about individuals.

Improving people's living standards is accompanied by the expansion of resource consumption. This is the contradiction between the "Libertarian City" and the "Republican City". If Chinese people's per capita living space, water, and electricity consumption were aligned with the US per capita standard, there might not be enough resources in the entire world. Therefore, on the one hand, the standard of living must be constantly improved, as a liberal demand. On the other hand, the consumption of material resources cannot be expanded and must even be reduced, for example, achieving carbon neutrality as the demand of a Republican City. This is impossible to achieve without the optimization of material resources by machine intelligence, for example, the optimization of road resources. The City Brain is the technological solution to reconcile the set of contradictions between material resource consumption and living standard improvement. The ideal of the City Brain is to "achieve the same quality of life of the people with 1/10 of the consumption of material materials". This is also a practical problem of the third wave architects oriented to the digital society and digital economy.

Case 2: NICE2035, Designing the Program

Buildings serve as a physical embodiment of culture, but their physical form cannot iterate as rapidly as their digital counterparts. NICE2035 aims to modernize and improve existing physical urban spaces through digital morphologies and serve as a model for larger-scale urban renewal. The initiative was launched by Professor Lou Yongqi of Tongji University in a new worker district from the 1970s on Siping Road, Yangpu District, Shanghai. With a degree in urban planning, Professor Lou has worked for over a decade to separate "design" from architecture and urban planning education, creating an interdisciplinary profession that bridges different fields. Under his guidance, the School of Design and Creativity at Tongji University has become the top-ranked design school in Asia according to QS rankings.

The residential area of the new worker district has been transformed into a diverse community with the commercialization of real estate. However, there is now a lack of connection between subdistrict residents. In contrast, commercial houses from the 1990s onwards used new forms of organizations, such as owners' committees and community properties, to provide living governance functions. In neighborhoods around Siping Road, the typical manifestation is the lack of coordination and management.

Professor Lou's team has restored the ground floor commercial area of the neighborhood, transforming it into a variety of urban functions. These "archetype labs" generate various proposals for future living scenarios and test them based on the community scale. The term "archetype" combines both schema design and iteration: it is not only a form but also a design with substance and a system that constantly optimizes and updates itself based on context. In physical urban spaces, it is difficult to change the function of neighborhoods and the physical hybridity of functions created by social divisions. However, digital programs facilitate the uninstallation, connecting the online community to the offline physical space through the "localization of the network," as well as connecting individual people and different public spheres. Each "program" defines its own protocol, which generates user behavior and interaction.

The "archetypes" are transmitted and duplicated by means of stories. They are spread through official media such as *People's Daily*, *Xinmin Evening News*, and *The Standard News*, as well as commercial and digital media. Not every archetype is successful, but they all contain iterations and updates that bring more mobility to the city and allow different people and events to intersect in time and space.

Newcomers to the community are not content with just participating in community discussions through Web2. They are actively experimenting with community governance as a "shared city." NICE2035 is developing a program called "The Decentralized Agent Profile Network," which

回报。政府在数据安全和人工智能伦理上承担着不可或缺的监管和立法工作，而新的技术手段（如"联邦学习"技术）也有可能实现不含有个体隐私信息的数据价值的交换，而不进行含有隐私信息的数据本身的交换。

需要指出的是，整体和个体的博弈一直发生在城市规划和设计的讨论中。个体生活水平提高的同时，面对的是巨大的整体资源消耗。如果中国人均居住面积、用水用电量等和美国标准对齐，那么将远远超过地球能提供的资源。因此，在有限的资源水平下提高生活水平，就需要对资源的使用效率进行优化。城市大脑虽然是一种技术手段，但是却实现了建筑师一直讨论的时空关系设计（比如：20世纪60年代到70年代的那些乌托邦城市提案，例如Archigram的行走城市、Cedric Price的Fun Palace等），从而有可能对空间使用进行动态的终极优化，实现"用1/10的物质资料消耗，实现同样的人民生活质量"，这是王坚博士提出的目标，不论1/10还是9/10，都是第三浪建筑师要面对的实践问题。

案例二：NICE2035的程序原型设计

建筑是文化的物质载体，物质形态无法像数字形态那样快速迭代。2035生活原型街（简称：NICE 2035）是通过数字形态对现存物质形态进行迭代的原型尝试。NICE 2035由同济大学娄永琪教授发起，他毕业于城市规划专业，过去十几年都致力于让"设计学"从建筑与规划的体系中独立出来成为连接不同领域的跨学科专业。他所领导的同济大学设计创意学院在短短的十多年历史中，一跃成为QS亚洲排名第一的设计学院。NICE 2035坐落在上海市杨浦区四平路一个典型的70年代工人新区。随着时间变迁，工人新区逐渐发展成一个多元居民小区。原来工人新区的居民通过"单位"把工作和生活联系在一起，当"单位"让位于市场化的时候，这些工人新区缺少了纯市场化商品房的物业配套等对应的设施，在四平路的典型表现就是小区内共同区域缺少协调和管理，商业价值和服务价值都逐渐衰退。

娄永琪教授的团队把小区中经营欠佳的底层商业区域承接下来，引入多元的城市功能，如当代首饰与新文化中心、同济—麻省理工上海城市科学实验室、朱哲琴声音实验室、Fablab创客工坊、同济—阿斯顿马丁创意实验室、同济阿普塔未来包装实验室等空间。这些被称为"原型实验室"的群落产出未来生活场景的各种原型并基于社区进行测试，通过社区——这一直面问题、贴近消费、多样且高容错率的城市分区——来重新启发如何设计城市。比起城市规划，"城市原型"兼具了模式设计和模式迭代：原型就像程序一样不停更新。在实体的城市空间中，我们很难改变城市街区的功能，物理上的混合功能以及自由主义城市形成的社会分化造成重重困难。而数字化的程序则方便卸载，将线上的社区和线下的物理空间建立关联，也让人的个体和不同公共之间产生关联。程序和所在的小区之间的关系就像"协议（Protocol）"，对介入小区的行为进行规定，例如实体空间中，不能随地吐痰、需要排队等，在线上空间则是善意、分享信息等。每个"程序"都会定义自己的协议，从而产生用户的行为和交互方式。[6]

原型是不需要被绘制的，因此任何图纸、模型、效果图等经典再现方式似乎都是徒劳的。原型用故事的方式来传播和复制，既有官方的《人民日报》《新民晚报》《文汇报》等，又有商业媒体和数字媒体，如"一条"记录了在NICE2035生活和工作的意大利国宝设计师Aldo Cibic的故事，带来了超过100万的阅读量。原型并不都是成功的，就像程序要不断更新和替代那样。每次去NICE2035，都会看到有人站在Aston Martin实验室前张望，这个实验室也鲜有使用，也许是时间更新了。

新加入这个社区的人并不满意只是通过小区微信群来解决家常问题和进行团购，而是希望可以进行主动社

NICE2035
NICE2035现场

will record contributions to the community through Web3 technology. Community members will be able to earn digital badges that record the "value recognition" of the community. This includes participation in activities, helping others, and identity verification. This enables real-time identification and recording of individual values within a collective that were previously difficult to quantify. When individual badges are distributed everywhere to form a badge network, the intersection of multiple actions becomes a network of social relationships between people. The distributed and untamable nature of Web3 technology provides a natural credible basis for this network, where actors seek reliable collaboration partners and interactions, and these actions bring new nodes and drive the expansion of the network, resulting in a tighter and more cohesive social network.

Case 3: The Practice of Action, Systematic Technical Capital

It's widely recognized in Silicon Valley that design can improve the integration of user experience and technology (according to John Maeda's Design in Tech Report). As a result, many tech entrepreneurs with design and architecture backgrounds, such as Airbnb, Pinterest, and Wework, have founded successful companies. In China, there are also many entrepreneurs with architecture and design backgrounds. For instance, Ren Zhengfei, the founder of Huawei, actually graduated with a major in architecture. However, the architectural entrepreneurs discussed here follow the entrepreneurial trajectory of Silicon Valley, focusing on innovation in products, technologies, and business models, which are typical characteristics of technological entrepreneurship. Prof. WANG Min, the creative director of Beijing Olympic Games, Prof. LOU Yongqi, the dean of Design and Creativity at Tongji University School, and Prof. TONG Huiming from Guangzhou Academy of Fine Arts School of Design are all enthusiastic supporters of design entrepreneurship in China. They believe that designers should use entrepreneurship to promote the interdisciplinary development of design further. In architectural practice, acquiring both capital and technology is often necessary, as designing a building requires a financial backer and a team of technically skilled engineers and constructors. Architects often face the "expediency" or "gamble" of balancing design choices and budget constraints.

However, entrepreneurship provides a way to overcome this "expediency" by acquiring capital and implementing technology simultaneously. Among numerous examples, four are noteworthy: Xkool, founded by architect HE Wanyu, who studied at the Belgrade Institute in the Netherlands and worked for Koolhaas at OMA (the name of her firm, Xkool, echoes their relationship). Xkool is an application of artificial intelligence in architectural design that improves the working efficiency of architects. While digital tools are frequently used in architectural design, Xkool aims to reduce friction in the design process and make it smoother through artificial intelligence and digital technology. MeetBest, founded by architect HE Yong, who studied at Columbia University in the US, is a shared activity space venture that connects idle activity spaces with distinctive features through an online booking system. There are many event spaces in the city with prominent design features, and the use of activity spaces is highly time-sensitive. MeetBest expands the time dimension of the event space. Modelo, founded by architect SU Qi, who studied at Harvard University and the University of Southern California, wants to lower the threshold of displaying and collaborating on 3D space and models so that it will be as easy to use 3D models as collaborating on documents and writing code. Modelo is used by many domestic and foreign design offices and has been acquired by "Kujiale," known as the Autodesk of China. Tezign, founded by architect FAN Ling, who studied at Princeton University and Harvard University and taught at the University of California, Berkeley, began as a community that combined technology and design; the firm has since evolved into the largest design production and management platform in China, serving over two hundred large companies, including Unilever and Starbucks, and fifty thousand design teams.

One of the prototypes located in the Siping Road community, ASTON MARTIN laboratory
坐落在四平路社区中的原型之一，ASTON MARTIN 实验室

区治理尝试。NICE 2035 正在开发一个名为"去中心化行动者网络"的程序，把社区贡献通过区块链技术进行记录，社区成员获得数字徽章（包括参与活动的认证、帮助他人的评价、身份认证等）。这使过去难以量化的个人在集体中的价值得以实时认定和记录。更重要的是，当分布在各处的单个徽章形成了徽章网络，多方的行动交集显化为人与人之间的社会关系网络。区块链技术所提供的分布式、不可篡改特性为该网络带来天然的可信基础，行动者就此寻找可靠的协作伙伴和寻求互动，这些行动带来新的节点与联系并推动着网络的扩展，形成更为紧密和高粘度的社会关系网络。

案例三：系统性实践和技术资本

硅谷的普遍认知是设计可以把用户体验和技术更好的结合，涌现了一批设计和建筑背景的科技创业者成立的公司，如 Airbnb、Pinterest、Wework 等。在中国，建筑和设计背景的创业者并不少见（很多人可能没有想到：华为创始人任正非先生就是学建筑学的）。但是这里所讲的建筑师创业者是指延续了硅谷的创业轨迹，具有典型的技术创业特点，以产品、技术、商业模式的创新为核心。设计学者王敏教授、娄永琪教授、童慧明教授等都是国内设计创业的坚定推动者。他们认为，设计师应该用创业这种手段，进一步推动设计的跨学科发展和整个体系的建立。与此相对的是"权宜建筑"，典型的建筑学创作往往受制于资本和技术，因为建筑设计一定需要有出钱的甲方、配合技术的工程师和工程施工团队。我们常常听到的是建筑师的"权宜"，建筑师是系统的一部分，而无法改变系统全体。[7]

创业是突破"权宜"的方式：一方面自己获取资本，另一方面自己实现技术。有四个值得关注的建筑师的创业实践包括：1. 小库（Xkool）是由何婉瑜创立，何婉瑜曾在荷兰的贝尔拉格学院求学，并在库哈斯的 OMA 工作（小库的名字也回应了这个关系）。小库是一个人工智能的建筑设计应用，可以提高建筑师工作的效率。建筑设计中往往将数字工具作为新形式生成的方式，小库却希望通过人工智能和数字化技术减少设计过程中的摩擦，让设计流程更顺畅。2. 好处（MeetBest）由何勇创立，何曾在美国哥伦比亚大学求学。好处是一个共享活动空间的创业项目，把有特色的闲散活动空间通过网络预订系统连接起来。城市中大量活动空间有显著的设计特色，而活动空间的使用又具有很强的时效性。好处拓展了活动空间在时间上延展的维度。3. 模袋（Modelo）由苏麒创立，苏麒曾在美国哈佛大学和南加州大学求学。模袋希望把三维空间和模型的展示、协作的门槛降低，可以像协作文档、协作写代码一样简单的使用三维模型。模袋被不少国内外的设计事务所使用，并被有中国的 Autodesk 之称的"酷家乐"公司收购。4. 特赞（Tezign）由范凌创立，范凌曾在美国普林斯顿大学和哈佛大学求学，并在加州大学伯克利分校任教。特赞起初是一个把科技和设计结合的社区，后来逐步发展成为中国最大的设计内容生产和管理的平台，服务包括联合利华、星巴克等在内的超过 200 家大型企业和 5 万个设计团队。

这些建筑师作为创业者试图用行动来回应社会和城市的问题，他们虽然具有很强的建筑学基因，但不满意只通过物质形式解决问题，而是希望建立系统，并运用资本和技术作为影响力的杠杆来解决问题。这种企图在建筑学的实验历史上并不陌生，包豪斯通过运用新的材料和技术，发明新的风格和类型，推动社会改良。公共住宅类型的出现，强调公民的社会性；逐渐减少装饰，采用钢筋混凝土，提倡机械美学，强调大众化等。这些设计创业者的意愿也推动了数字时代的社会发展和改良，这个初心和他们从事建筑学习和实践的时候一样，但是实践的环境从物理地块变为了系统。对他们来说，系统不是设计的概念，而就是设计的对象本身。他们不是通

These architects - entrepreneurs are seeking to address social and urban issues through their initiatives. With their strong architectural background, they are not content with solutions that are purely physical in nature, but instead aim to create systems that leverage technology and capital for greater impact. This kind of experimentation is not new to the field of architecture, as we can see in the example of the Bauhaus movement, which sought to drive social progress through the invention of new styles, typologies, materials, and technologies. The emergence of public housing typologies that emphasized citizenship's social nature led to reduced ornamentation, the use of reinforced concrete, the promotion of mechanical aesthetics, and a focus on massing, among other changes. Today, design entrepreneurs' goal remains the same: to promote social development and progress in the digital age, with their focus shifting from physical sites to digital systems. For these architects, the system itself is the object of their design, rather than just a conceptual tool. They don't simply express their "attitude" through physical projects, but instead reveal it through their business, which serves as an ongoing mission, vision, values, and dynamic object that is constantly evolving. Although they are no longer solely architects, they continue to "design and construct" urban media.

Conclusion

The third wave of architecture assumes that architecture is an action (i.e., a structure) rather than the consequence of a certain paradigm. This action is meant to harmonize the relationship of four contextual archetypes (authoritarian, libertarian, republican, and communitarian) with people. On one hand, authority structures are governing and controlling individuals with unprecedented efficiency (e.g., the urban brain). On the other hand, certain urban model (e.g., NICE2035) have effectively enhanced the attributes of the shared city, bringing political and public life back into the public sphere, and enabling a richer and more active "structuring" of urban relations through virtual forms such as WeChat groups, purchase groups, and even Decentralized Autonomous Organizations (DAOs). In contrast, the communitarian city emphasizes a culture of sharing, reminiscent of the neighborhood relationships of the past, where physical and social distances were consistent. However, in the contemporary context where cities are growing larger and the structure of the urban population is increasingly segmented, this shared ideal often becomes wishful thinking that is overly formalized without supporting modes and instruments of governance. The attempts of the architects of the first and second waves to restore the shared community have not been effective but have either resulted in the shared form exclusively occupied by individuals in the Libertarian city (e.g., "Commune at the foot of the Great Wall") or the shared ideal imposed on the liberty of individuals (e.g., "Vanke Tulou"). The architects who emerged from the third wave of architecture as entrepreneurs have more distinctly incorporated technology, capital, and business modes into the scope of their own design. It is worth reiterating that third-wave architecture is a hypothesis for analysis instead of a conclusion. The reason why this discussion is meaningful is because of the invariable architectural primacy and the dramatic transformation of various means of production-arithmetic, program, capital, and creativity.

过做项目来表达自己的"态度"，而是通过创立和经营企业——一个具有持之以恒的使命、愿景、价值观，并不断自我迭代、充满活力的客体来进行实践。他们已经不是建筑师的身份，但却在持续地"架构"着系统。

总结

第三浪建筑学的假设是：建筑学是一个动作（即：架构），而不仅是形式。一方面，机器智能正在史无前例的高效地管理、调整、优化每一个个体（如城市大脑）；另一个方面，有些城市原型（如 NICE 2035 等）让城市的共有性得以回归，公共生活又回到了（数字的）公共场域中，而且通过微信群、团购群，甚至去中心化组织（DAO）等虚拟形态实现更丰富和主动的"架构"城市关系。相比较而言，共有性强调分享，就像过去城镇生活中的邻里关系，物理的距离和社会关系的距离是一致的。但是在当代城市不断变大、城市人群结构不断分化的现状下，共有的理想常常变的一厢情愿，显得过于形式化，治理模式和工具上却没有进行配套。第一浪和第二浪的建筑师们的聚落形式尝试并没有有效地还原社区和共有的状态，其结果要么成为自由主义城市下被个体独享共有形式（如"长城脚下的公社"），要么共有理想有可能被强加于个体自由之上（如"万科土楼"）。第三浪建筑师们只有一个作品，这个作品就是毕生之作（Oeuvre），更加鲜明的把数字技术、资本和商业模式纳入自己可以设计和迭代的范畴之内。值得再次重申的是，第三浪建筑学只是一种分析的假设，而并不是结论。而之所以认为这个讨论具有意义，是因为不变的建筑学初心和正在形成中的各种生产资料——算力、程序、资本和创意。

参考文献：
[1] Toffler, Alvin. The Third Wave: The Classic Study of Tomorrow. New York: Bantam, 2022.
[2] 曾鸣. 智能商业 [M]. 北京：中信出版社 ,2018.
[3] Architecting 一词来自美国学者 Molly Steenson 的说法，具体可参见：Architectural intelligence: how designers, tinkerers, and architects created the digital landscape. Cambridge, MA: MIT press, 2017. 中文翻译成"架构"来自于计算机领域中把 Architecture、Architect 分别翻译成"架构"和"架构师"。
[4] Christopher Alexander, Cedric Price, and Nicholas Negroponte and MIT's Architecture Machine Group. 2014. Architectures of information [M]. NJ: Princeton University.
[5] 王坚. 在线：数据改变商业本质，计算重塑经济未来 [M]. 北京：中信出版社 ,2016.
[6] 把城市和程序、平台、协议等网络结构进行比较，最早来自于荷兰数字人文学者 Martijn de Waal，具体内容可参见：Waal, Martijn de. 2014. The City as Interface: How Digital Media Are Changing the City. Rotterdam: nai010 publishers.
[7] 李翔宁. 权宜建筑——青年建筑师与中国策略. 时代建筑，2005 (6), 16-21.11. QIN X, YANG T. Infinite space within the cube: Synaesthetic City[J]. Urban Design, 2020(2): 14-23.

FAN Ling Tongji University Design A.I. Lab. Director, Professor, Doctoral Supervisor
CHEN Xueer Tongji University Design A.I. Lab. Postgraduate

范凌 同济大学设计人工智能实验室主任、教授、博士生导师
陈雪儿 同济大学设计人工智能实验室硕士研究生

基金项目：本文为 2020 年度教育部人文社会科学研究规划基金项目资助的"人工智能语境下的艺术设计实践趋势研究"（编号：20YJA760011）的研究成果。

Media and Beyond
不止媒介

李涵 胡妍
LI HAN HU YAN

尹毓俊
YIN YUJUN

冯果川
FENG GUOCHUAN

The Complete Map of Capital Beijing 京师全图
Drawing Architecture Studio 绘造社   (pp. 184–189)

The Grand Stage 大戏台
Drawing Architecture Studio 绘造社   (pp. 190–195)

Garden of Fantasies 幻园
Yushanfankuan 鱼山饭宽   (pp. 196–203)

Spatial instigation: village, factory, city
空间策动：村，厂，市（以研究与展览作为空间策动的工具）
Atelier Alternative Architecture 多样建筑   (pp. 204–225)

唐煜
TANG YU

NEXTMIXING—The Spatial Experiments on Usage
那行——使用的空间实验
NEXTMIXING 那行   (pp. 226–237)

张烁
ZHANG SHUO

Archild 童筑文化
Archild 童筑文化   (pp. 238–241)

曾仁臻
ZENG RENZHEN

Shituzi Architecture Channel 使徒子建筑频道
Zidao Culture 子道文化   (pp. 242–243)

覃清峡
QIN QINGXIA

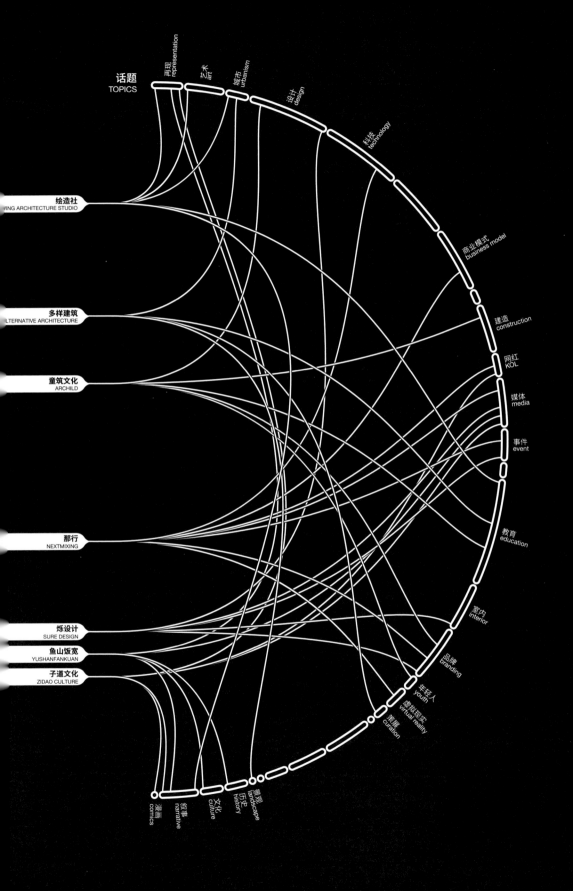

Dialogue:
Media and Beyond

LI Han / HU Yan / ZENG Renzhen / YIN Yujun / TANG Yu / FENG Guochuan / QIN Qingxia / LI Jingjun

对话：
不止媒介
李涵 / 胡妍 / 曾仁臻 / 尹毓俊 / 唐煜 / 冯果川 / 覃清硖 / 李憬君

WANG Fei: All of you are practicing or have practiced architecture with multiple identities, such as artist, cartoonist, curator, space operator, media person, KOL, and children's educator. This group is closely related to media and representation. Could you please briefly introduce your understanding of social media and media? Let's start with the three experts on new media: TANG Yu (Archmixing | Nextmixing), QIN Qingxia (Zidao Culture), and LI Jingjun (LI Lexian Media).

TANG Yu: If we consider media as a communication platform like an official account, then I have a very close relationship with the media, whether it's "Archmixing," "Nextmixing," or the "Parallel Space" that I am working on. From an architect's point of view, I think media is a closed loop on the operational level of an architectural company. In reality, it is very difficult for everyone to visit the buildings after the construction work is completed, but the architecture firm can post their articles, videos, and photos on different media, which is also a way to promote their architectural brands and ideas. Therefore, I think this is a "closed loop" state from the perspective of the operating firm.

From the standpoint of the "Yi Tiao" consultant, I regard the medium as a fantastic way of perception and communication. This is actually a fortuitous opportunity. When "Yi Tiao" approached me, I realized that many people, including government staff and owners of some large enterprises, were not aware of the capabilities of outstanding young and middle-aged domestic architects. In contrast, they probably had more contact with foreign star firms or large design institutes. Therefore, with the development of network technology from 3G to 4G and the rise of short videos in 2014, there came wonderful opportunities to promote China's young and middle-aged architects. Initially, I intended to promote the entire Chinese architectural industry on a cultural level at "Yi Tiao." For the general public, this kind of promotion helps them become more acquainted with these outstanding domestic architects. In fact, the architectural practice and theoretical ability of domestic architects are on par with those of foreign architects. Therefore, in this sense, I think media is a way of perception and communication.

QIN Qingxia: Mr. TANG's ideas align perfectly with my own. Firstly, the concept of the "closed loop" is crucial. Throughout the years, I have spent most of my time creating comics and engaging in self-media, while also dabbling in landscape and architectural design. However, after investing tens of millions or even hundreds of millions of dollars in a project, party A may lose interest in it once it's completed. By integrating media into the project, we can boost its influence and ensure that it is not forgotten. This is similar to how car manufacturers market their new models by holding promotional shows and placing advertisements on various media platforms. Thus, I am currently working to establish this closed-loop concept in the field of architecture.

Additionally, I began systematically creating architectural content for self-media platforms last year. Previously, I sporadically posted articles

王飞： 大家都曾经或者正在从事建筑实践，很多都是注册建筑师，大家目前也都有着多重设计的身份，艺术家、漫画家、策展人、空间运营家、媒体人、流量大咖、儿童教育家等等，我们这一组都与媒体、媒介有紧密的关联，请各位简短介绍一下自己对当今社会媒体、媒介的认识。我们先从三位新媒体的专家，唐煜（阿克米星｜那行）、覃清硖（子道文化）、李憬君（李乐贤文化媒体）开始。

唐煜： 假如把媒介看成"公众号"一样的传播平台，那么无论是在阿科米星和那行文化，还是正在做的"平行空间"，我和媒介的关系都是非常紧密的。如果是站在一个建筑师的角度来看，我觉得媒介是建筑公司运营层面的一种闭环。现实中，这些公司的作品建成后，大众很难人人都去现场参观，但建筑公司可以把关于作品的文章、视频和照片利用不同媒介传播出去，这也是完成了对其建筑品牌以及思想的推广。因此，从运营事务所层面来讲，我认为这是一种"闭环"状态。

如果是站在"一条"的顾问的立场上来看，我觉得媒介是一种很好的认知与沟通的方式。这其实是一个非常机缘巧合的事情。当"一条"公司来找我的时候，我认为当时国人对于我们本土的一些优秀中青年建筑师的能力是不太熟知的，包括政府或者是一些大型企业的业主。相比之下他们可能更多接触的是国外的明星事务所或者大型的设计院。因此，2014年随着网络科技从3G发展成4G以及短视频的兴起，我觉得这是一个推广国内中青年建筑师很好的机遇。我在"一条"的初衷是在文化层面上推广整个中国建筑界。这种推广对于普罗大众来说，是帮助他们认识并了解中国这些优秀的本土建筑师。事实上，国内建筑师的建筑实践与理论能力是完全不亚于国外的建筑师的。因此，从这点意义上来讲我觉得媒介是一种认知与沟通的方式。

覃清硖： 我觉得唐老师说的挺对的，很多东西也跟我想的比较像。首先就是"闭环"的这个概念。这么多年，我大部分的时间是放在漫画以及一些自媒体内容上，但我也在做一些景观设计和建筑设计。我觉得有时候甲方可能投资了好几千万甚至上亿的项目，但做完了就放在那里了无人问津。其实媒体是可以成为整个项目的一部分，来帮助它传播，增加它的影响力。类似汽车领域，当车厂出了新的车型，他们会有对应的营销手段，较大的品牌如果要批量的上新，甚至会举办推广的秀场，投放不同的媒体，来帮助他们的宣传。所以我在做的事情一方面就是在建筑领域建立这个闭环的概念。另一方面是我一直在做自媒体，我从去年开始系统的做建筑类的内容，之前会零零星星的发一点微博或者写个公众号文章，但是没有系统性，也不像现在这样每周都有更新视频。我觉得之前是没有找到一个合适的视频的切入点，

Qin Qingxia's Architects Media Series in Bilibili © Zidao Culture
棉仓城市客厅，常州 © 阿科米星

Qin Qingxia's Architects Media Series in Bilibili © Zidao Culture
覃清硤的 bilibili 网站建筑师视频系列 © 子道文化

on Weibo and public accounts without any weekly updates. Initially, I struggled to find a suitable entry point for videos, but after watching knowledge-based videos on BiliBili App made by "Half Buddha" and "Little John Khan," I decided to create architecture-themed videos myself. The response was great, so I continued to promote this approach.

I aspire to represent the construction industry on self-media platforms. In the past, social and entertainment news dominated Weibo, and there was no systematic platform to promote architecture and aesthetics. New media is similar to bringing offline experiences to the online world. For example, I used to give lectures on university campuses or in public spaces, but now many lectures are delivered online. Therefore, I believe that the line between virtual and real and traditional and new media is becoming increasingly blurred. In the future, with the emergence of the Metaverse era, many activities may be carried out online.

LI Jingjun: The new media that I am referring to is similar to what TANG Yu and QIN Qingxia are doing now. We want to convey our knowledge of architecture to the public through media. However, after stepping into this field, I realized how insular we were.

In my view, I use Zhihu as a platform to bridge the gap between professional architectural papers and public articles. In professional media, we write many papers in an obscure and incomprehensible way, which makes it challenging for the public to understand. But when it comes to the internet platform, we should explain things in simple language to help them understand from a public perception perspective.

At the same time, Zhihu is strategically optimized on Baidu search. If you search for common questions on Baidu, the first two are typically advertisements, and the third is Zhihu. So, from my perspective, Zhihu is a more professional version of Baidu. The platform aims to answer users' queries, and we take this goal into account in our content production operations. In my later operations, I also started to build a team. The LI Lexian account looks like a personal account, but it is operated by a team. Over the past few years, we have produced more than 700,000 words of answers and content. Although there are fewer high-quality questions nowadays, our high-quality answers usually occupy the top positions for most architecture-related questions.

I have my own understanding of self-media operation. On one hand, there must be a personal point of view, also known as a "KOL" (key opinion leader). This is very critical, but where does this opinion come from? Through one's cognitive improvement or personal experience, one produces their own ideas, and then refines and shares these ideas. On the other hand, I think the platform is also very important; one needs the platform to support them. In a highly competitive market, good content doesn't always come out on top, as the content that catches the public's attention is often not so good. If we want to present something really cool and helpful to the public, we need platform support.

Finally, I would like to explain why I am probably going to be the biggest architecture blogger on Zhihu. The first thing I want to say is that I became a Big V before good respondents joined this platform. In the early days, content creation was free, but most people still expected to make a profit. As a content creator, I didn't make any profit for at least seven years, no advertising, and it was not easy to stick with it for so long. I think the second important aspect is to have some

但随着所有的自媒体都在做视频化，我也看到bilibili网站出现了好多新的知识类的视频，比如"半佛仙人"，"小约翰可汗"的那种形式的视频，我觉得特别适合用来做建筑主题，所以我当时就尝试着去自己做这个东西，后来看看效果确实也还不错，所以就继续地推进这个事情。

我是很希望能在这个自媒体平台上为我们的建筑行业去寻到一些发声的机会，过去微博上大部分都是一些娱乐新闻以及社会新闻，有关建筑以及美学的新闻都是非常碎片化的，没有一个系统化的平台去推广建筑。

我觉得新媒体有点像是把我们过去的一些线下的场景去搬到线上，比如过去我可能是在大学的校园里讲课，或者受邀去一些公开场合做讲座等，但是现在很多东西都变成线上的了。因此我觉得虚拟跟现实的边界越来越模糊，传统媒体和这种新媒体的边界也在模糊，我们未来可能出现的元宇宙时代，也许很多事物也是线上进行的。

李憬君：我所说的新媒体可能跟唐老师、使徒子覃学长正在做的事情很接近，其实是想把我们专业的知识传递给大众，因为其实做了自媒体以后才发现我们这个圈子有多闭塞。

从我的角度上面讲，它有很多时候是一种关注度的匮乏，我可能以知乎为平台填补了专业的建筑论文和普罗大众之间的那一点认知的偏差。比如说在专业的媒体上面，其实像我们写很多论文，用句晦涩难懂，不说

人话的方式，给大众一种难以理解，故弄玄虚的感觉。但是到了这种公域互联网的平台上，就是要从流量或者是从他们的认知考虑，以公众认知范围内的角度作为一个切入点，深入浅出才能真的帮助他们理解到这个事情。

大家现在都叫知乎小百度，因为每个到知乎上面来提问的人或者来看内容的人，他们都是带着问题来的，所以目的指向性非常强。同时知乎平台也是在百度搜索平台上做了战略优化。如果你们在百度里面去搜索一些比较常见的问题，一般前面两个是广告，第三个就是知乎，所以从我的角度上面来讲，知乎就是一个更加专业的百度。因为知乎平台试着回应这些目的性的，我们在内容生产的把控上也会把这种目的性纳入考虑的过程中。在我后来的运营中，也开始组建团队，李乐贤这个账号看上去是个人账号，但其实是由一个团队运营的。通过过去这几年的布局，我们差不多布局了70多万字的回答跟内容。虽然现在鲜有高质量问题的提出，但我们比较高质量的回答基本上能占据大部分建筑相关问题排名靠前的位置。通过知乎这一平台，我也认识了很多新的朋友，另一方面很重要的也让更多的平台找到了我。

关于自媒体运营的经验心得方面，我觉得第一，一定要有个人观点，也称为"意见领袖"：key opinion leader（KOL）。这个是很关键的，那么这个opinion又是从哪里来？是通过了自己的认知的提升，或者是自己个人经历，才能产生自己的想法，然后把这个想法提炼

Glenindiv bespoke men's custom made suite shop, designed and v co-owned by Li Jingjun

李憬君联合创立并设计的高端定制男装店 Glenindiv bespoke

Li Lexian Media's KOL architecture series media in Bilibili.

bilibili 网站上李乐贤媒体的建筑自媒体视频

175

insights of your own. In addition, we have now become a matrix model to put excellent content creators together, and then we can organize events, allowing fans to know each other. Thus, for a platform to increase its followers, building a stable community is essential. People join a new content platform for the community, not just for a particular creator. If only a few specific creators are active, and only a few people are creating content, the platform is unlikely to develop. The growth of a platform requires the power of a cluster.

WANG Fei: YIN Yujun (Atelier of Alternative Architecture), and LI Han and HU Yan (Drawing Architecture Studio), as curators and designers who have participated in many exhibitions, what do you think is the gap between the professional audience and the general public? How can you balance academic, public, and business interests?

YIN Yujun: I have been reflecting on our journey to where we are today. After much contemplation, I realized that our development process is closely tied to Shenzhen. I believe that the city has presented us with many opportunities. Additionally, our personal interests, such as organizing exhibitions, have also played a significant role. As we don't teach at universities, we are less likely to engage in academic research projects. Therefore, I created my own path, and being a curator for the Shenzhen Biennale has allowed me the freedom to develop my ideas. As a result, since we started curating the

Shenzhen Biennale of Architecture and Urbanism, Guangming Sub-venue, curated by YIN Yujun, 2017 © Atelier Alternative Architecture

尹毓俊策展的深圳建筑城市双年展光明分展场，2017 年 © 多样建筑

sub-venue, we always leave a foreshadowing of each exhibition, which I call the "curators' special section." This section is reserved for us to explore the academic research behind the year's theme and our interests, in order to broaden its depth and breadth. This approach allows us to invite suitable exhibitors and academic institutions to participate in the exhibition.

The exhibition usually lasts for a year. We spend most of our time doing curatorial work and creating special sections before inviting exhibitors. By doing this, we satisfy our research status need. Additionally, exhibitions provide a closed loop that allows us to design scenarios where the house can be used. Every time we organize an exhibition, we are essentially designing situations where the house is used or at least exhibited in various contexts. Through this approach, we observe several scenarios in which the house is used, and our office can gradually move towards a more integrated state.

Many times when we start a project, it is not the beginning of an architectural design. It may be for planning, exhibition, or even writing a script for some government department. After this, we get involved in the planning of the theme exhibition, which is also related to the house, and finally, we design the building. All these processes are linked together slowly. So, we architects are no longer just a part of the ring, but we may start from the beginning.

In response to Professor WANG Fei's question, most of the exhibitions we organize are professional exhibitions, such as exhibitions and architecture exhibitions in Shenzhen. However, I believe that exhibitions must include two parts: one is its professionalism, which requires a solid foundation of knowledge production; the other is the follow-up publication after the exhibition, the process of turning knowledge production into actual knowledge, which is also a kind of professional experience. At the same time, I think the exhibition should present a large number of elements that can be accepted and understood by the public in order to have better public communication. However, there is always a powerful barrier when it comes to achieving this. I am not quite capable

并分享出来，做一个进一步的提升。然后另外一方面，我觉得平台同样也很重要，一定要有一个支持你的平台，因为现在大环境很卷，好的内容经常是出不来，为什么？因为夺大众眼球的内容往往不是好的内容，我们想要做一些真的是有干货的、真的是对大部分人有帮助的内容，这其实也是需要一些平台的支持。

最后说说为什么我可能马上就能成为知乎最大的一个建筑类的博主。从我的角度上面来讲，可能一方面是入场比较早，平台还没有完全收集到很多不错的回答者的比较早的时候就拿到了大V。早期的内容创作都是无偿的，而那时大部分人还是比较希望可以看到一个收益的，但是作为一个内容创作者来说的话，我至少前7年是没有任何收益的，也没有广告，想要坚持这么久也不容易。第二个我觉得比较重要的一方面，还是要有一些干货和自己的一个见解。另外一个我们现在也变成了一个矩阵化的模式，我们把这些优秀内容的作者都给聚到一起，然后大家可以去做一些梦幻联动，那么在这样子一个情况下，粉丝之间也会相互去做一些关注。那么对于一个平台的涨粉，或者在平台上面形成一个稳定的社群也是非常重要的。因为大家加入一个新的内容平台不可能是因为某个特定的内容创作者，如果只有某个或某几个特定的创作者在发声，只有他一个人在做这个内容的话，这个板块是无法发展的。板块的成长需要一个集群的力量，所以类似的平台也有一部分的社群的属性。

王飞： 尹毓俊（多样建筑）、李涵和胡妍（绘造社），作为策展人和经常参与展览的设计师，一方面是专业的观众，另一方面是非专业的大众，这个度在哪里？在学术、大众、商业之间如何平衡？

尹毓俊： 我一直也在想，到底是通过什么样的路走到今天呢？我思考了很久以后发现其实发展的过程主要跟深圳有关，我觉得是深圳带来了很多机遇。另外我觉得可能是跟个人的兴趣有关系，比如说做展览这件事，其实我们每次做项目都是带有一定的私心的，我自己对研究展览是比较感兴趣的，但是在中国研究这个事情缺乏土壤：没有基金会，也就没有所谓的资金支持。因为我们也不在大学里面长期任教，所以基本上我们也不太可能在大学里去做类似科学研究的项目。所以我自己就发现了一个路径：深圳双年展这个平台它提供了一定限度的自由，可以发展自己想做的项目。当变成了一个策展人以后，可以做的事情就变得更多，因为这个展览基本上是以我的想法去安排的。

所以从开始做分展场的策展人以后，每一次展览我们都会留下一个伏笔，这个伏笔我称之为策展人特别单元，这种单元是自己留一个私心。我们针对今年的主题和自己的兴趣点去发掘背后的学术研究，从而更好地帮助这个展览去拓宽它的深度以及广度，有了这样的一个整体的研究以后，我们才能去邀请适合这个展览的参展人和一些学术机构。

所以一般这个展览持续可能一年的时间，可能我们用大半年的时间先来做策展人特别单元，然后后续才陆续安排参展人。这样的一个情况下，其实是满足了我们自己的所谓的研究的状态。另外一件事情就是，做展览这件事情就是一个闭环。我觉得年轻这一代建筑师可能都会好奇设计的房子在后来是如何被使用的，也会好奇所谓的建筑师想象的场景能不能在其中被实现。所以对于我们来说，每次做一个展览的话，其实是设计这个房子被使用的情况，至少它在这个房子被展览的时候，可以实现其中的一个场景。

换句话说，通过这样的一个方式，其实我们是可以观察到这个房子它真正被使用的几种情况，在这样的一种情况下，我们现在的事务所可能慢慢的走向了一个比较综合的状态。我们可能很多时候切入一个项目时，它不是作为一个建筑设计开始，它可能是针对策划或者展览，甚至是可能给一些政府部门写剧本。然后先有了这个东西，才开始介入策划可能需要的开幕的主题展，这种开幕的主题展也会跟这个房子有关系，最后才着手开始设计这样一个房子。慢慢这些事情和过程会串联在一起。所以我们觉得这件事情就像我们建筑师不再是这当中的某一个部分，而是我们可能从一开始就在做这件事。

所以说回到王飞教授的这个问题的话，我们做大部分的展览其实都是专业展览，比如说深圳的展览和建筑展。但是我觉得展览一定会构成两个部分，一个部分就是它的专业性，也就是说我刚才所说的知识生产的部分，它需要有一个完整的知识生产，以及展览结束以后，它有一个后续的出版物，把知识生产变成真的知识的过程，这也是一个专业性的体验。但同时我们也得有大量能被

of making exhibitions more accessible to the public in a popular way. Perhaps we are currently limited in terms of the dissemination of the exhibition or the so-called more popular state, but we are slowly gaining better understanding through contacts with different projects.

LI Han: Our debut was quite different from that of most professional architects. While most architects start with designing buildings, we began our journey with an illustrated book called *A Little Bit of Beijing*. We consider traditional media, such as books and magazines, to be our primary media. Even though audio and video have since emerged, as well as new media based on the internet, we chose the most traditional media form of all, paper books. We wanted to emulate "Archigram," who began their work with magazines in the 60s and 70s. So even though we may seem more focused on media communication than other architects, we are still quite traditional.

Our way of thinking may differ somewhat from that of most practicing architects. Despite the lack of complete practice work, we have had many opportunities to exhibit our work because of this difference. The advantage of exhibiting is that it allows us more freedom in terms of the content of our work. This is similar to what YIN Yujun mentioned earlier, using exhibitions as an opportunity for research or thematic thinking. Our usual work is similar to architects who accept commissions for designing, except we accept commissions for drawing. Our clients may have many commercial considerations for their requests, so it can be challenging for us to elaborate on these commissioned works in terms of ideas or academics.

There are also different types of exhibitions. For example, in the Venice Biennale we participated in 2018, the curator, Dr. LI Xiangning, asked us to paint Taobao Village. At first, I had no idea what Taobao Village was, but the topic proved to be a worthwhile subject for in-depth research using painting as a medium. When we completed the work "Taobao Village - Half Acre City," others who were concerned about the phenomenon of Taobao Village referenced or cited our painting in their research or discussions. This was, to some extent, an affirmation of our way of conducting urban research through painting.

Another category of exhibitions is represented by the Shenzhen Biennale, which may have broader themes. For example, the 2017 Shenzhen Biennale was held in an urban village, and the theme was "Urban Symbiosis." We happened to be observing a "hole-in-the-wall" in the building on Dirty Street in Sanlitun, which was demolished, and this became a significant problem in the discussion of urban symbiosis. We hadn't considered presenting this issue in the form of a painting, but with the theme of the exhibition, we were motivated to conduct a more in-depth research and to push forward to complete the project. Otherwise, this valuable topic might have remained on hold without a chance to take shape.

HU Yan: I believe we are situated in the middle of the spectrum between mass and professional dimensions, and that there are various gradients in between. We might not be able to comprehend certain content that is at the highest level of professionalism, but at the same time, we cannot cater to the needs of the public without any conditions. Therefore, we adjust our position back and forth on this gradient according to different cases. When we design exhibitions for others or exhibit our works, we consider the habits of today's mass media or general audience. For instance, we cannot ignore the public's desire to take photos and check-in. In my opinion, it is inappropriate to focus only on our professionalism without considering the public's demand for punch card moments at exhibitions. This approach is not beneficial to the industry's promotion at the public level. As creators, we desire our work to be appreciated by more people, so we think we should embrace changes as much as possible while still considering how much we should embrace.

However, the prerequisite for all this is still to concentrate on professional quality, and the content should at least reach a standard that satisfies the creator. Moreover, it is ideal to have a good topic that can attract public attention. We attempt to avoid highly abstract and deep content

Dongbei Renaissance, Shenyang, 2021 © Drawing Architecture Studio
东北文艺复兴，沈阳，2021 © 绘造社

大众接受和理解的部分，其实也是为了让这个展览有更好的大众传播度和理解，但是我觉得这件事情一直存在一个很强的壁垒。对于我来说，我不太有能力把这件事情用比较大众的方式让大家更好去理解，可能我们现在相对来说比较局限的方面是展览的传播度，或者说所谓的更大众的状态，我们也是慢慢接触不同的项目后有了更好的理解。

李涵：我们出道的定位，可能跟职业建筑师是不一样的。大多数建筑师靠建筑作品出道，而我们是靠创作《一点儿北京》的绘本走上这条道路的。当时我们理解的传统媒介就是书和杂志，当然那是最老的媒介，即使后来又有音频、视频，到现在基于互联网的一切新媒体，我们选择的媒介还是最传统的，那就是纸质的书。当时我们想模仿的对象是电讯派，他们在20世纪六七十年代以杂志为媒介开始创作。我们也想这样起步。所以尽管看上去我们跟其他建筑师比好像是更特别、更偏向媒介传播一点，但其实也是非常传统的。

可能我们的想法跟大多数实践建筑师有点儿不同。我们也因为这样的不同获得很多参展的机会，尽管我们没有什么建成的实践作品。而参展的好处，是展览对作品内容的要求相对比较自由。我觉得这比较像尹毓俊刚才提到的，把参展这件事作为一个做研究的督促，或者是一个进行专题性思考的机会。我们平时的工作内容也跟建筑师接受委托做设计一样，我们是接受委托来画张画儿。而客户对画儿的要求可能有很多商业上的考虑，我们很难从思想上或者学术方面去阐述这些委托作品。

展览也有不同的类型。比如我们在2018年参加的威尼斯双年展，创作题目是命题作文，策展人李翔宁老师让我们画淘宝村。在他第一次提到淘宝村时，我甚至都不知道那是什么。虽然这是一个命题作文，但这个题目是一个非常值得讨论的话题，值得我们去以绘画作为媒介进行深入研究。在完成《淘宝村·半亩城》这个作品后，也有其他一些关注淘宝村现象的人在他们的研究讨论中参考或引用这幅画，这也是一定程度上对于我们通过绘画来进行城市研究的模式的肯定。

另一类展览以深圳双年展为代表，命题可能是个比较宽泛的主题。比如说2017年的深圳双年展的举办场地是城中村，主题叫"城市共生"。正好那年我们一直观察的三里屯脏街上的那栋楼里的"开墙打洞"被拆除了，这刚好也是讨论城市共生时特别大的一个难题。本来我们也没想过如果以绘画形式展现这个问题会得到什么结

and seek out topics that are relevant to ordinary people. Nonetheless, sometimes the understanding of popular topics may not necessarily be universal among the general public. So, we embrace popular topics from a professional point of view rather than the other way around because we have our own professional standards, and this is also the bottom line for our content production. This approach is similar to the idea of self-publishing content production and question-answering that was mentioned by LI Jingjun earlier. We strive to keep up with current trends while maintaining our professional standards, and if we cannot keep up, we will go with the flow.

WANG Fei: ZENG Renzhen (Yushanfankuan), while everyone is shifting towards non-conventional media, you have always been sticking to conventional media such as ink and rice paper with both historical and contemporary content. What is your original aspiration, and how do you view non-conventional media?

ZENG Renzhen: Initially, my paintings were created as a preparation for the construction of gardens. I consider painting as a type of research related to landscape gardening, and in my paintings, I try to blend ancient scenes with modern life, aiming to bridge the gap between tradition and modernity. To express this idea, I utilize brushes and rice paper, which aligns with the traditions of Chinese gardening and landscape painting. By inheriting the tradition of medium, I attempt to create something new while respecting the past.

My understanding of the medium is still relatively limited, but I believe that it is primarily concerned with expression and communication. For me, the most important thing is to create my work, and how it is presented or transformed is less crucial. Recently, I've been watching old movies by King Hu, a director whose martial arts films from the 1960s and 1970s have greatly influenced contemporary directors like Ang Lee and Tsui Hark. Despite not knowing martial arts, he used Peking Opera as inspiration for the martial arts action in his movies. By borrowing from traditional stage action, he was able to create something new. Similarly, I use the oldest tools of brushes, ink, and paper to create something new while remaining rooted in tradition. For me, using a brush and paper is a habit developed since childhood, and as a Chinese, it is a natural way for me to express myself.

While some people may embrace new technologies and methods of communication, it is not necessary to question the traditional methods used by others to express their thoughts. Some people, like myself, find it easier to express thoughts through traditional methods rather than digital medium. Ultimately, I believe my work is new, like new wine in old bottles, and the core thinking in my work is unique. Therefore, I am not concerned with the means or technology used to spread my ideas.

WANG Fei: FENG Guochuan (Archild), as you have been educating young children for many years, could you tell us about your original aspirations, experiences, and future prospects in this field?

FENG Guochuan: Rather than the architecture itself, I care about people. Through the discipline of architecture, I can observe the relationship between humans and their environment. As trained architects, we understand that architecture can serve as a "medium" that influences living conditions and the relationship between people. Architecture can not only discipline and monitor people, but also activate and even liberate them. My goal is to promote and influence people's awareness through discussions about architecture. Architecture allows people to become aware of how their subjectivity is shaped by the environment.

The most visible changes in media over the past decade have been the development of self-media and the birth of the paid knowledge mode in the internet era, which have had a serious impact on both traditional media and architecture. The new social interactions that were once supported by architectural space are now increasingly detached from architecture. A simple and small live-streaming room can easily replace our grand lecture halls, and more importantly, the number

果，但有了这个展览题目之后，我们也有了进行深入研究的督促，有了动力推进完成这个项目。要不然这么宝贵的题目可能会一直被搁置，没有机会成型。

胡妍：我觉得我们处于大众和专业维度的中间地带。从大众到专业之间是有很多梯度的，比如说可能我们也无法理解一些在专业性上位于最高层面的内容，但也无法去无条件迎合一般大众所需求的内容。我们也在这个梯度上根据不同的个案来回调整自己的定位。我们的参展作品，或者为别人设计的展览，肯定会考虑到当今大众媒体或者普通观众看展的习惯，比如你不能忽略掉大家要拍照打卡这种需求。如果这时候只顾着发散自己的专业性而不回应大众对于展览打卡的需求，在我看来是非常不合适的，而且这也不利于这个行业在大众层面的推广。作为一个创作者，你肯定希望你的作品能被更多人欣赏，所以我们觉得应该尽可能拥抱变化，当然也要判断拥抱的程度是多少。

当然，这一切的前提还是要注重专业的质量，内容起码要达到令自己满意的标准。在此基础上，如果能有一个好的话题性，能吸引大众关注则再好不过。我们也会尽量避免一些特别深奥抽象的内容，寻找更多和普通人息息相关的话题，即使有时候你对大众话题的理解可能并不一定在大众群体中具有普遍性。所以我们是从专业角度出发，拥抱大众话题，而不会反其道而行之，因为我们的作品有我们自己的专业标准，这也是我们对自己的内容生产的底线。这其实和李乐贤刚刚介绍的做自媒体内容生产和问题回答的思路很相似，我们在保持自己专业标准的情况下试着抓住当下潮流，如果实在跟不上，也就顺其自然了。

王飞：曾仁臻（鱼山饭宽），当大家都在走向非传统媒体的时候，你一直在坚持传统媒体，水墨宣纸，内容既有历史，也有现代，有什么初衷及展望？对非传统媒体有何看法？

曾仁臻：我画画的初衷是为造园子做准备，绘画是作为一种山水园林有关的研究，绘画内容会兼顾古今，有古代的情景也有现代的生活，是想弥合传统与现代之间的裂痕。以毛笔宣纸进行创作，表达诗情画意，是符合中国山水园林、山水画一直以来的媒介传统的。我继承了这个媒介的传统，使用最古老的工具，是希望专注做好内容的更新和突破。

我对媒介的理解还是比较粗浅的，我理解的第一个点就是"表达"。比如说我的方式就是以绘画的方式表达。所以按我的理解来说，我最关心的还是自己所思所想，首先要自己做出作品来，至于之后作品以什么方式传递和以什么形式转化，都不是我应该去思考太多的问题。所以我最近在看老的电影，看胡金铨六七十年代拍的武侠片，胡金铨是我最近最喜欢的一个导演，同时它影响了李安和徐克等大导演。胡金铨他不会武术，他也不知道如何用真实的武术动作的方式去拍电影，但他熟悉京剧，他找的武术指导是京剧演员，电影里的武打实际是借用了京剧上的舞台动作，运用的是京剧的肢体语言或眼神的方式，你就发现他居然用了一个非常传统的表意方式去做电影这种全新的创作。换句话说，我也是用毛笔、用最传统的方式在创作，对我来说笔墨纸砚就是我的母语。作为一个中国人，我用毛笔写毛笔字是一种习惯，我从小就是这么个习惯。你不能说现在有英语所以你用中文就是不对的，所以从我的角度来说，可能很多人会拥抱新的东西，拥抱新的一种社会的交流方式，但不用质疑别人所用的传统方式，传统也是活的东西，有的人

Huan Yuan (Fantacy Garden) by ZENG Renzhen
曾仁臻（鱼山饭宽）的《幻园》

ARCHILD's design-build workshop, founded and taught by FENG Guochuan

冯果川创立并教授的童筑文化课程

of viewers has increased significantly. The venue of the live broadcast is very spontaneous, flexible, and variable, much like in a nomadic state. In contrast, our buildings on the ground are high-cost, time-consuming, and immovable once they are built. In the past, architecture used to be an intermediary between people and space. Now, the relationship between people and space has become more diverse and flexible. The era of architecture monopolizing the relationship between people and space is over, but the value of architecture as a discipline for thinking about the relationship between people and space is even more important. That is why I have shifted the focus of my work from designing buildings to discussing and disseminating architectural knowledge, and the object of my dialogue has also shifted from architectural professionals to non-professionals, especially children.

I believe that what I do is not professional enlightenment, but rather a kind of quality education that every ordinary person needs. Before teaching children how to design, I first teach them to observe their own lives. For example, I ask children to observe whether each family member has a corresponding piece of furniture or to see where they all like to stay. Some children find that their mothers are always busy in the kitchen, while their fathers are motionless as if they have grown together with the sofa. We also found that there was one family that was particularly harmonious.

They liked to stay in bed when they were free. The most communal space in a typical family is the sofa in the living room, where everyone may sit on one side of the sofa. However, this girl had a family of three who inclined to staying on the bed, and her mom liked to put her feet on her. Her dad liked playing games, while her mom often read magazines, and she just kept playing there. The family's bodies were interlaced on the bed, which reminded me of Robin Evans's depiction of Raphael's painting characteristics. Raphael was always keen to depict warm, soft bodies gathered together. I felt that the boundaries of their bodies were blurred, and the intimacy of this gathering was something that could not be brought by the sofa. In contrast, the bed was the furniture that could create more possibilities for interaction between bodies, and I was particularly touched by that.

I believe that my purpose in educating children is to encourage them to focus on people's living conditions. Children are less stereotyped than adults, and it is easier to interact and communicate with them. With a few years of educational training, children can acquire knowledge in this area, regardless of whether they pursue architecture as a career in the future. Education for children needs to take into account their specific developmental period, and architecture classes can help to awaken or respond to their instinctive impulse to build while also encouraging their parents to learn alongside them.

可能用的更加得心应手，比如我想表达什么就能用它表达出来，反而我不知道该怎么用电脑图绘板来表达我的思考。但是我认为我的作品是新的，大概就是老瓶装新酒的意思。因为作品中核心的思考是崭新的，所以我不在乎传播它的手段和技术。

王飞：冯果川（童筑文化），作为深入培养未来的希望、祖国的花朵的园丁们，你们对教育儿童有什么初衷、心得和展望？

冯果川：我并不关心建筑，我关心的是人，通过建筑学我看到的是人与空间的关系，看到的是人生活的状态。作为建筑师，我们受过的训练让我们理解建筑作为一种"媒介"可以调整人的生活状态，可以调节人跟人之间的关系，建筑既可以规训和监视人，也可以激活甚至解放人。我希望通过思考和讨论建筑这种媒介，促进和影响人的状态。建筑学让大家意识到我们在这个环境中的主体性是如何被塑造的。

这十几年里最显著的媒体的变化是在移动互联时代下的自媒体的发展和知识付费模式的诞生，这既是对传统媒体的严重冲击，其实也是对建筑学的严重冲击，因为这些新的互动和社交形式曾经是依靠建筑空间来支撑的，现在越来越脱离建筑了。简单狭小的直播间就可以取代我们的千人报告厅，而且观看人数更多。直播的地点很随意，场景灵活更变，可以说直播的人处在一种游牧状态。而我们的建筑却昂贵、建造缓慢，建成后就笨拙地呆在原地不动。原来建筑是把人和空间结合在一起的中介，现在人和空间的关系变得更多样、更灵活，建筑垄断人与空间关系的时代结束了，但是建筑学作为思考人与空间关系的学科价值却还在，甚至更加重要了。所以我把自己工作的重心从设计建筑物，转向建筑学知识的讨论和传播，选择对话的对象也从原来的建筑专业人士转向了非专业人士，特别是儿童。

我觉得我做的不是职业启蒙，而是一种每个普通人都需要的素质教育。在教孩子们设计之前，我先教他们如何观察自己的生活。例如，我让孩子们分析每一个家庭成员有没有一件对应的家具，或者看他们都喜欢待在哪个地方，有的小朋友发现母亲总是在厨房忙得团团转，而父亲却和沙发像长在了一起似的一动不动。我们也发现有一家人特别融洽，他们没事就一家人都呆在床上。一般家庭的最有公共性的空间都是客厅的沙发，大家可能各坐沙发一边。但这个女孩却是一家三口待在床上，她妈妈还喜欢把脚架在女儿的身上。她爸爸就一直在打游戏，她妈妈就一直在看杂志，她就一直在那玩。他们一家人的身体是在床上交错在一起，让我想到罗宾·埃文斯对拉斐尔绘画特征的捕捉，拉斐尔总是热衷于描绘聚集在一起的温暖而柔软的身体。我觉得他们身体的边界是很模糊的，这种聚集的亲密的关系确实是沙发无法满足的，相比之下床才是更能创造身体之间互动可能性的家具，我当时就特别的感动。

我觉得我做儿童教育的目的就是带领孩子们去关注人的生活状态。相比起成年人，儿童的成见不是很重，也更容易和他们有所互动和交流。孩子们通过几年的教育培养，他们就会有这方面的一些观察和认识，之后至于他们是否会学习建筑，这并不重要。儿童教育也是抓住孩子们成长的特定时期，建筑课唤醒或回应了儿童渴望建造的本能冲动。这也能带动孩子们的父母一起学习。

Drawing Architecture Studio
The Complete Map of Capital Beijing
Beijing, China 2021

绘造社
京师全图
中国，北京 2021

The Complete Map of Capital Beijing is a contemporary replica of the original map from the Qing Dynasty. Spanning an area of 700 × 700 meters, it features a blend of some of the most iconic architecture in Beijing: from the Yong He Gong Lama Temple of the Qing Dynasty to the Chinese Anglican Church built during the Republican Era; from the socialist mansion An Hua Lou around the founding period of PRC to the residential areas developed in the 1980s and 1990s; and from the traditional hutongs in the inner city to today's urban villages. It serves as a nostalgic memory, exploration, and imaginative journey through Beijing.

This exhibition showcases *The Complete Map of Capital Beijing* in the form of an architectural diorama, featuring three-dimensional models from two-dimensional images. These models were created using style transfer algorithms that combined classic prototypes in the history of Western modern architecture with the amateur language of Beijing vernacular constructions.

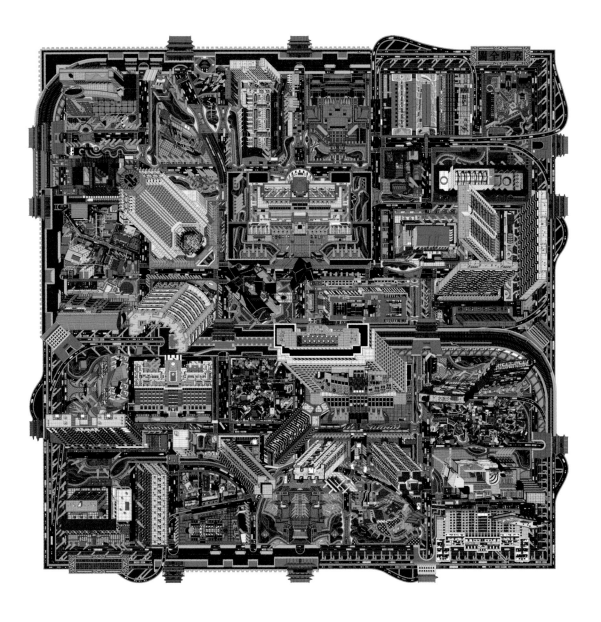

pp.184-185: The Complete Map of Capital Beijing exhibited in Wind H Art Center, Beijing (2021), © WANG Hongyue.
This page, above: The Complete Map of Capital Beijing 2D drawing. This page, below: Details of diaroma of The Complete Map of Capital Beijing. Opposite: The Complete Map of Capital Beijing exhibited in Wind H Art Center, Beijing (2021).

第 184-185 页：《京师全图》在北京山中天艺术中心展出（2021 年），© 王洪跃。
本页，上：《京师全图》二维图纸。本页，下：《京师全图》图纸与模型局部放大。对页：《京师全图》在北京山中天艺术中心展出（2021 年）。

Through personal reinterpretation, these models take on a strange yet familiar form in Capital Beijing.

The creation process of *The Complete Map of Capital Beijing* is a self-organized "game." Drawing Architecture Studio provides the framework and creative path of Capital Beijing, while participants (students from Syracuse University School of Architecture) freely fill in the framework while being guided by a series of provisos on "style and language." *The Complete Map of Capital Beijing* encourages a cooperative attitude and actively seeks more participants, allowing for accidents, compromises, and overdoing, but at the same time, embracing unity of command to ensure that the collective can form a joint force to continuously advance on the track, rather than going round in circles and canceling out each other.

The particularity of architecture requires architects to think through intermediaries, such as drawings and models. Although dealing with intermediaries takes up most of the design time, architects often overlook the self-generated independence of those intermediaries in the pursuit of ultimate architecture. *The Complete Map of Capital Beijing* is an adventure of architectural drawings and models. In this adventure, architects no longer envision the giant buildings through intermediaries but strive to transform them into independent "selves" and use these "selves" to build an "independent world" on the architectural diorama.

Credits and Data
Project Title: The Complete Map of Capital Beijing
Year of Completion: 2021
Location: Beijing, China
Project Team: (Drawing Architecture Studio) LI Han / HU Yan / ZHANG Xintong / ZHANG Yuanbo / LIU Ping
(Syracuse Architecture)
Drawing and Model: LUO Chenhao / HUANG Deqiang / ZHU Haihui / ZHONG Junye / ZHUANG Kaicheng / WANG Kexin / FENG Wenting / MENG Xinqi / TANG Xinyu / FENG Yiqun / LIU Yian / WANG Yuxuan / ZHANG Yi / ZHANG Yaqi / YANG Zhexu / ZHENG Zhi
Installation: XU Xiaoxuan / LYU Nuo / SUN Danlin / MA Jiazi / LYU Zizhou / ZHAO Chendong / YE Rulin / WANG Jiaqi / XIE Mingrui

《京师全图》是清代《京师全图》的当代摹本。在 700 米 ×700 米的范围内，拼凑杂糅混合了北京城的代表性建筑：从清代的寺庙雍和宫到民国时期的中华圣公会教堂；从建国初期的社会主义大厦安化楼到 20 世纪 80 年代至 90 年代的住宅小区；从传统的内城胡同到当代的城中村。它是关于北京的回忆、漫游和想象。

《京师全图》以建筑沙盘的形式呈现，三维的建筑模型从二维的图像中生长起来。这些建筑模型用图像迁移算法混合了西方现代建筑史中的经典原型和北京民间建造的业余语言，并通过个人的再次解读，以一种既陌生又熟悉的形式降临京师。

《京师全图》的创作过程是一场自组织的"游戏"。绘造社提供了京师的框架和创作路径。参与者（雪城大学建筑学院学生）在框架的内部自由填充，同时又被一系列硬性规定的"风格语言"限制。《京师全图》呼唤一种协同劳动的心态，积极吸纳更多的参与者；开放地接受意外、妥协和过度用力；但同时拥抱统一指挥，以保证集体可以形成合力沿着路径不断推进，而不是原地打转，相互抵消。

建筑的特殊性让建筑师必须通过中间媒介——模型与绘图来思考。尽管与中间媒介打交道占据了设计的大部分时间，但对于终极建筑的渴望让建筑师往往忽略了中间媒介自生的独立属性。《京师全图》是一场建筑图纸和模型的冒险。建筑师们在这次冒险中不再通过中介展望巨大的建筑，而是努力让它们成为独立的"自我"，并通过这些"自我"构建一个沙盘上的"独立世界"。

Opposite: The Complete Map of Capital Beijing exhibited in Wind H Art Center, Beijing (2021). This page: The Complete Map of Capital Beijing installed in the entrance of E-Park, Beijing (2022).

对页：《京师全图》在北京山中天艺术中心展出（2021 年）。本页：《京师全图》在北京 E-Park 主入口陈列（2022 年）。

Drawing Architecture Studio
The Grand Stage
Foshan, China 2022

绘造社
大戏台
中国，佛山 2022

Next to the road, there is a parking building backed by Xiqiao Mountain and facing the thousand-acre mulberry fish pond, creating a striking yet fitting contrast. While its modern architectural style and infrastructure seem incompatible with the natural surroundings, the building's substantial volume evokes a sense of commemoration. On a larger scale, it echoes the grandeur of Xiqiao Mountain and the vast water surface of the fish pond, acting as a human-made node in dialogue with the surrounding environment.

The Grand Stage, constructed using scaffolding and Oxford cloth, transforms the south facade of the parking building into a large-scale performance stage facing the fish pond. Foshan, the birthplace of Cantonese Opera, is honored by this stage, paying tribute to the traditional culture while celebrating the grand stage of life in a broader sense. The center of *the Grand Stage* features a landscape of Xiqiao town, composed of colored cloth collages that depict a mix of modern and traditional buildings such as ancestral halls, presenting a dynamic scroll of daily life. Along the stairs, visitors can climb up to the raised stage, becoming actors in the drama and enjoying the stunning view of the fish pond. Despite its imposing size, the softness and ever-changing folds of the cloth give the stage a sense of lightness and flexibility, creating a dialogue with nature through different expressions with the change of wind and rain.

The Grand Stage is both a monument and a billboard, symbolizing traditional culture and serving as a metaphor for contemporary cities. It inherits the internal contradictions of the parking lot, and while trying to reconcile them, it also produces new paradoxes. It is an artistic spectacle that is both reasonable and absurd.

pp.190-191: transform the south facade of the parking building into a super-scale big stage, © TIAN Fangfang. This page, above: Elevation of Xiqiao Mountain South Gate Parking Building, © Drawing Architecture Studio. This page, below: First and final design sketches for The Grand Stage.

第 190-191 页：将停车楼南立面转化为一个超尺度的大戏台，© 田方方。
本页，上：西樵山南门停车楼立面，© 绘造社。本页，下：《大戏台》设计草图初稿和终稿。

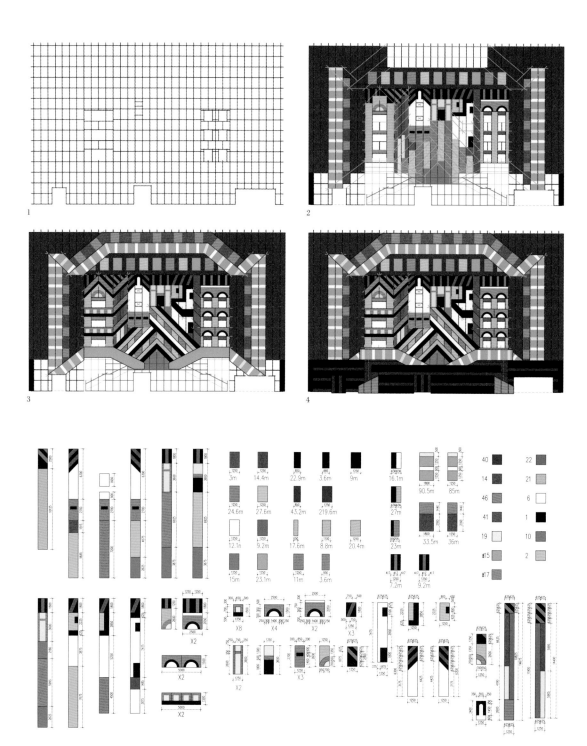

This page, above: A layering design for the installation of The Grand Stage. 1.1st layer of scaffolding 2.1st layer of fabrics + 2nd layer of scaffolding 3.2nd layer of fabrics 4. Staircases covered with fabrics. This page, below: First and final design sketches for The Grand Stage.

本页，上：《大戏台》分层安装设计图。1. 第一层脚手架　2. 第一层布料＋第二层脚手架　3. 第二层布料　4. 覆盖布料的台阶。本页，下：《大戏台》布料图案制作设计图。

一座停车楼矗立在公路旁,它背靠西樵山,面向千亩桑基鱼塘,显得既突兀又恰当。一方面它的现代建筑风格和基础设施与周边的自然山水格格不入,但另一方面它巨大的体量又形成了某种纪念性。从更大的地景尺度来看,它与雄伟的西樵山和广阔的鱼塘水面相呼应,成为与周围环境对话的人造节点。

《大戏台》用脚手架和牛津布将面向鱼塘的停车楼南立面转化为一个超尺度的大戏台。佛山是粤剧的发源地,《大戏台》既是对这一传统文化的致敬,也是在更大、更广的意义上向生活这座大戏台致敬。《大戏台》的中央是用彩布拼贴的西樵镇景观:现代住宅、传统民居和祠堂等建筑鳞次栉比,日常生活的画卷以一种戏剧化的方式呈现。顺着楼梯,参观者可以登上升高的戏台,他们在俯望鱼塘景观的同时,也成为舞台上的演员,进入到了戏剧的场景中。尽管尺度巨大,但是布匹的柔软性和千变万化的褶皱又赋予戏台这个庞然大物一种轻盈感和灵活性。它会随着风雨的变化呈现出不同的表情,从而与自然形成对话。

《大戏台》是一个纪念碑,也是一个广告牌。它是传统文化的象征,也是当代城市的隐喻。它继承了停车楼自身的矛盾性,在试图调和先天的矛盾的同时,又产生新的悖谬。它是一个合理又荒诞的艺术奇观。

Credits and Data
Project Title: The Grand Stage
Year of Completion: 2022
Location: Foshan, China
Project Team: LI Han / HU Yan / ZHANG Xintong / XIAO Junfu / DAI Kun

This page, above: Details of The Grand Stage. This page, below: Sewing and Installation process for the fabrics, © Drawing Architecture Studio. Opposite: The Grand Stage facing the fishpond with the mountain at its back.

本页，上：《大戏台》细节。本页，下：布料缝制与安装过程，© 绘造社。对页：背靠西樵山，面向鱼塘的《大戏台》。

  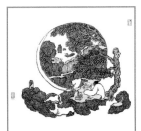 

pp. 196-197: Garden of Fantasies.
This page, above: Garden of Fantasies, Fantasy series, stage 1. This page, middle: Garden of Fantasies, Fantasy series, stage 2. This page, below: Garden of Fantasies, Fantasy series, stage 3. Opposite: Garden of Fantasies, Section series.

第 196-197 页：《幻园》。
本页，上：《幻园》幻系列第一阶段。本页，中：《幻园》幻系列第二阶段。本页，下：《幻园》幻系列第三阶段。对页：《幻园》剖系列。

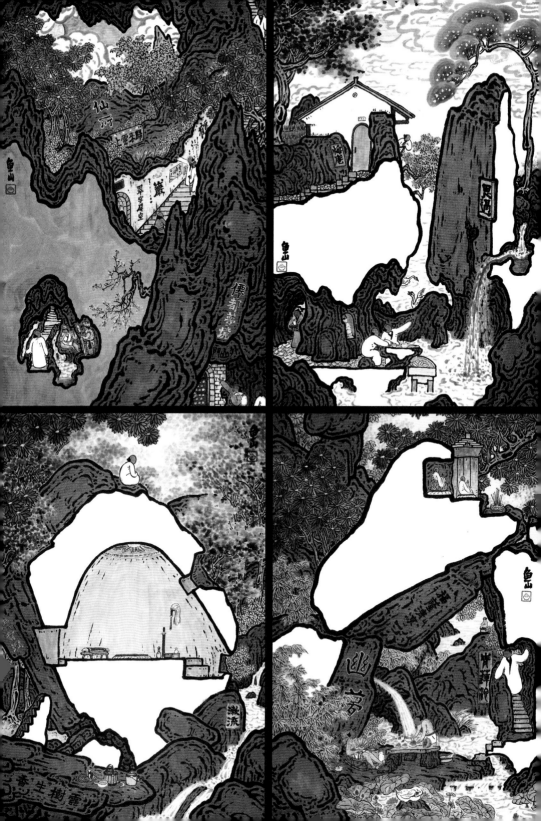

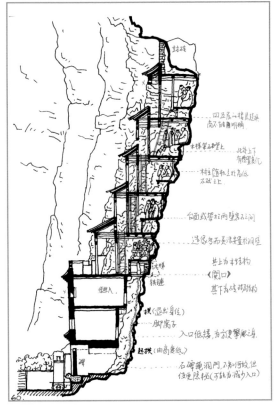
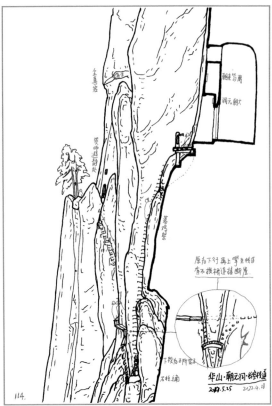

The creation of the *"Garden of Fantasies"* is based on the understanding and analysis of traditional Chinese painting, aiming to offer an alternative understanding of the relationship between "landscape painting-garden-architecture" through a more open and diverse perspective.

The main work, *"Fantasy Series,"* has advanced to its fourth stage. The first stage involves re-recognizing and reorganizing the wonders depicted in ancient landscape paintings. The second stage interweaves landscape and nature with architecture in a "half man-made and half natural" way, fostering new spatial relationships. The third stage is the gradual transformation and replacement of the natural elements in landscape paintings with architectural language, resulting in ideographic architecture that relates to the "habitable and travelable" aspects of the landscape. The fourth stage introduces the "profile" viewing method of landscape paintings. In this stage, we attempt to introduce the "profile" method of viewing into landscape paintings and reconsider the internal space and structure relationship within the landscape.

If the *"Fantasy Series"* presents a global, large-scale, and structural reflection, then the *"Red Series"* intends to portray courtyard scenes to supplement a detailed perception of daily routine in the close-scaled physical space. Meanwhile, the *"Green Series"* aims to discuss specifically the physicality of each group of people regarding the posture and position of trees and rocks, revealing possible sentiments between human and nature.

This page: Garden of Fantasies, Green series, stage 3. Opposite: design sketches.

本页:《幻园》绿系列。对页:手绘设计稿。

Credits and Data
Project title: Garden of Fantasies
Completion: 2014-2022
Painter: ZENG Renzhen (aka Yushanfankuan)
Media: Ink and paper

This page, Opposite: Garden of Fantasies, Red series.

本页，对页：《幻园》红系列。

《幻园》的绘画创作是基于对中国传统绘画媒介的理解与辨析，基于对"山水"这一文化脉络的观察与感受，尝试以更开放的思维形式与多样的视角来获取对"山水画——园林——建筑"这一组蒂联关系的新认知。

作为主体画作的"幻系列"已推进至第四阶段。第一阶段是对古代山水画中奇境的再识与重组；第二阶段

是按"一半人力,一半自然"的方式使山水自然与建筑交织一体,互动出新的空间关系;第三阶段山水自然部分渐渐被建筑的语言完全转化和替代,形成具备山水"可居可游"关系的表意建筑;第四阶段则开始尝试在山水画中引入"剖面"的观看方式,重新思考山水中的内部空间与构造关系。

如果说"幻系列"所呈现的思考是全局性的、大尺度的、结构性的,则"红系列"试图通过对庭院场景的刻画来补充对近身尺度空间内生活日常的细节认知,而"绿系列"更是要将身体性的讨论具体到每一组人与树石的姿态和位置关系,充分展现人与自然间可能生发的情趣。

Atelier Alternative Architecture
Spatial instigation: village, factory, city
Shenzhen, China 2017-Now

多样建筑
空间策动：村，厂，市（以研究与展览作为空间策动的工具）
中国，深圳 2013- 现在

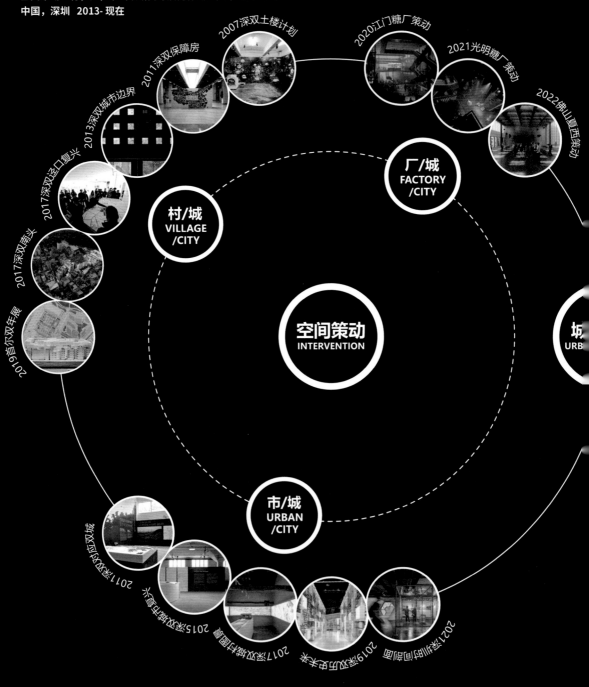

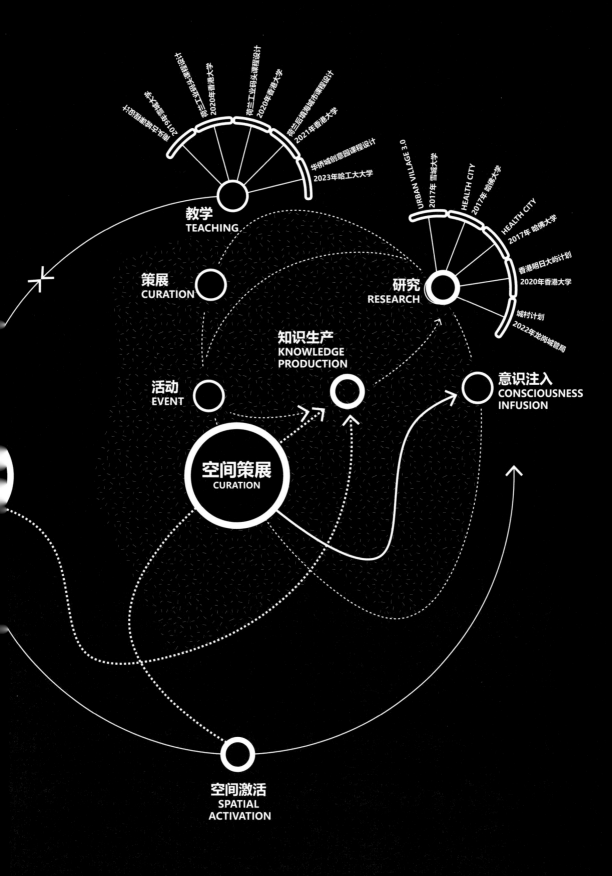

VILLAGE/CITY INSTIGATION-
城村激活

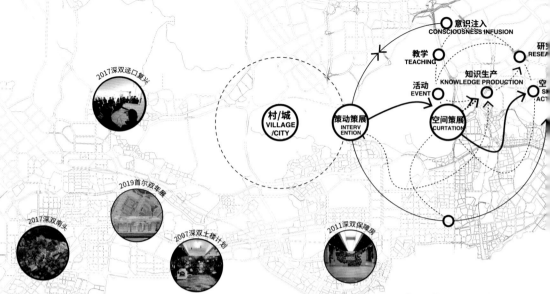

After undergoing several rounds of deep dual fermentation, in 2017, Nantou and Guangmingjingkou villages were selected to draw public attention towards urban villages through exhibitions. This initiative aimed at intervening in urban villages and forming a complete spatial planning plan with urban villages hosting exhibitions.

以深圳城中村为研究对象，探讨低收入居住议题，并以此为媒介产生白石洲、华为村等数届深双的展览和研究，以此策动大众对城中村的关注。经历数次深双发酵，2017年以南头和光明迳口村为起点，"以村为展，以展策村"构成完整的空间策动计划。

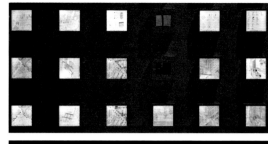

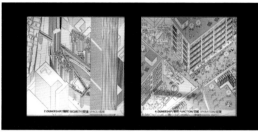

2013 UABB City Boundary—Baishizhou Research Exhibition

Urban village tenants, original residents, residents of commercial housing complexes, and outsiders. The encounter or juxtaposition of these people in space may trigger the presentation of these boundaries. Therefore, we attribute these border events to three types: property rights events, population events, and ethnic group events. "Implantation" is our design standpoint for this series of issues. We aim to intervene in this series of invisible boundaries using economical, flexible, and minimal designs, thereby causing a fundamental impact on them and weakening, eliminating, or improving their state.

2019 Seoul Biennale—multipilicty-Longhuahua Village Research and Exhibition

MultipliCity is an analysis and projection of labor housing in Shenzhen, contextualized within the urban village. The dormitories and villages surrounding Foxconn's Longhua campus serve as a case study, which raise a single question: what parameters can reflect the physical architecture of the urban village's social vibrancy and sense of community?

MultipliCity hypothesizes that the urban village can develop as a vibrant organism while preserving layers of physical and emotional history through air right negotiations between villagers, developers, and the government. MultipliCity imagines a future where the community grows upwards through parasitic structures, outcompeting the hyper-modernity of the globalized factory to create an adaptive social and physical interface for Chinese labor.

2013 深双城市边界——白石洲研究展览

城中村租户、原居民、商品房小区住户以及外来人员。在这些人群中，他们在空间中的"相遇"或"并置"，都会引发这些边界被呈现的可能。因而，我们把这些"边界"事件归结为三个类型：物权事件、人口事件和族群事件。"植入"，是我们对待这一系列问题的设计立场，我们希望能用一些经济、灵活与极小的设计去介入这一系列不可见的边界，对其产生"根本性的影响"，从而在成因上削弱、消除或改善边界的状态。

2019 首尔双年展城上之城——龙华华为村研究及展览

华为村是在城中村的背景下对深圳劳动力住房的分析和预测。以富士康龙华园区周围城中村为案例研究提出了一个问题：城中村的实体空间可以什么维度构成其社会活力和社区意识？

华为村假设通过村民、开发商和政府之间领空权的权谈判，城中村可以成长为一个充满活力的有机体，同时保留物质和情感历史的层次。华为村设想了一个社区通过寄生结构向上发展的未来，超越全球化工厂的超现代性，为中国劳动力创造一个适应性的社会和社区。

This page, left, Opposite: 2013 UABB City Boundary—Baishizhou Research Exhibition. This page, right: 2019 Seoul Biennale—multipilicty-Longhuahua Village Research and Exhibition

本页，右，对页：2013 深双城市边界——白石洲研究展览。本页，左：2019 首尔双年展城上之城——龙华华为村研究及展览。

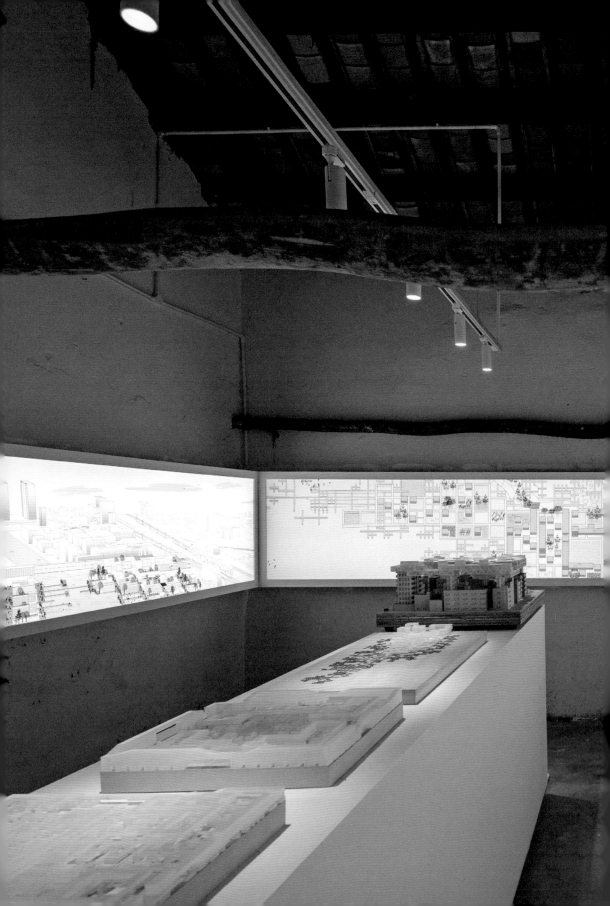

2017 深双迳口复兴

在经历数次城中村研究后，得以以村落改造+展览方式，对深圳的一个城中村进行整体空间激活

"迳口复兴！"通过展览的方式尝试寻找一种非经济导向的发展模式和激活社区的机会。在当下中国城市化进程中纯粹理论的讨论和实践之间，也许需要一种策略性的方式，来触碰理性的边界和政策条例的限定。因此，"迳口复兴！"不仅仅是一次展览，更是一次充满冒险的尝试，一次对当下未知的探索。

2017 UABB Guangming—Jingkou Revive!

"Jikou Revive!" is an exhibition where we aim to explore a non-economy oriented development model and create an opportunity to activate the community. In the current process of urbanization in China, a strategic approach may be necessary to navigate the boundaries of rationality and policy regulations, balancing theoretical discussions and practical applications. Therefore, "Jingkou Revive!" is not just an exhibition, but also an adventurous attempt, an exploration into the unknown.

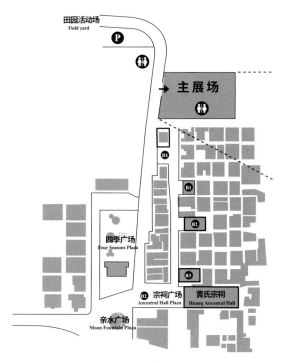

Qu project spatial structure

双年展空间介入结构

This page, left: jing aerial photography. This page, right: Comparison after the intervention of the Biennale. Opposite: Exhbition in the village.

本页，左：迳口村航拍。本页，右：双年展介入改造后。对页：村落展览。

The countryside's economic relationship and geopolitical differences distinguish it from the collective space of the city, and for a long time, it remained distant from the latter. The transfer of land functions, the outflow of young labor, and the continuous decline of economic income have changed the relationship between the countryside and the city from a coexistence and symbiotic one to an unbalanced status quo. Neo-Marxist scholar David Harvey captured this unbalanced phenomenon and placed it in the prism of capital production, where the countryside became a supply of raw materials and an output of labor in the chain of urban production relations, serving as the hinterland.

"Jikou Revive!" explores methods and models of community regeneration through field research, space activation, case comparisons, experience sharing, event introduction, and local cultural creation in the form of an exhibition. From the perspective of space activation, this exhibition provides a regeneration framework for creating new public living places in the community through the intervention of a series of public spaces. It aims to create a biennale and community regeneration with a bottom-up spatial organization and a top-down exhibition structure.

　　乡村作为一种因经济关系和地缘政治差异而区别于城市的集体空间,很长时间内与后者保持着距离。土地的功能流转,年轻劳动力的外流,经济收入不断下降使得乡村与城市的关系不再是一种共存共生的关系,而是不可避免地走向了一种不平衡的现状。这种不平衡的现象被新马克思主义的学者大卫·哈维所捕捉,并被放置于资本生产的棱镜中观察,乡村成为了城市生产关系链条中的劳动力输出和原材料提供的腹地。

　　"迳口复兴!"以展览的方式,通过实地研究、空间激活、案例对比、经验分享、活动引入及在地文创等多种方式重新探讨社区再生的方法与模式。从空间激活上,"通过一系列公共空间的介入,为社区的公共新生活场所营造提供了再生的框架。在此框架下,深双展览遍布迳口社区的公共广场、民居、展场和宗祠之中。本次展览是一次以自下而上的空间组织和自上而下的展览结构,形成双年展与社区再生的尝试。

This page, above: 2017 UABB Guangming invited unit. This page, below: Li Ning dancer and her team interact with the old village architectural scene of the body and dance performance. Opposite: Exhbition in the vellage.

本页,上:2017深双光明分展场邀请单元——中国乡村研究,城村关系研究。本页,下:李凝舞蹈家及其团队与老村落建筑场景互动的肢体与舞蹈表演。对页:村落展览。

URBAN/CITY INSTIGATION-
城市激活

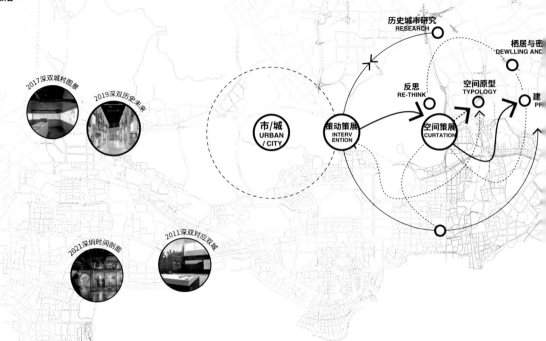

Focusing on the city as the research subject, we reflect on the urban development mode by investigating high-density living, public life, ecology, habitat, urban development history, and more. Our aim is to initiate research and exhibition to explore the possibility of spatial prototype and its role in guiding actual spatial design.

以城市为研究对象，从高密度居住、公共生活、生态、栖息地、城市发展历史等方面，反思城市发展的模式并以此策动研究与展览。在研究与展览之中，反思空间原型的可能性以及其如何以思想方式导引实际空间设计。

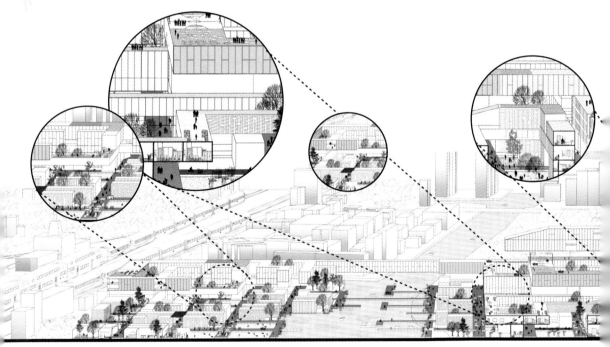

This page: 2017 UABB Village Landscape - RE-THINKING on Chinese-style living
本页：2017深双城村图景——反思中国式居住

A Panoramic View Of Urban-Rural Relations— Reflection On China's Urbanization Model

China's rapid urbanization over the past 30 years, driven by both state and private capital, has led to the convergence and uniformity of cities under the influence of modernist large-scale infrastructure construction and market-led utilitarian real estate development. Urban life has become simplified, resulting in reduced diversity. In this special exhibition unit, we propose four different project types, including villages, urban suburbs, cities, and urban villages, to study and reshape architectural types, spatial relations, life patterns, natural resources, economic relations, and social structures specific to each design. Through the exploration of urbanization in these different areas, we hope to carry out cross-scale research and design from urban design to architectural design.

城/村全景——中国城市化模型反思

中国的城市化在国家资本和私人资本的双重驱动下经历了三十多年的高速发展。在现代主义式大规模基础设施建设和市场主导的功利主义房地产开发下，城市的面貌和空间不可避免的走向趋同和单一。这也意味着城市的生活单一化和多样性的消减。策展人特别单元展通过村落、城市郊区、城市与城中村4个不同的项目提案，分别对建筑类型、空间关系、生活模式、自然资源、经济关系和社会结构以研究和重塑的方式，呈现于设计之中。通过这些不同地区的城市化可能性的探讨，展开从城市设计到建筑设计的一种跨尺度研究和设计。

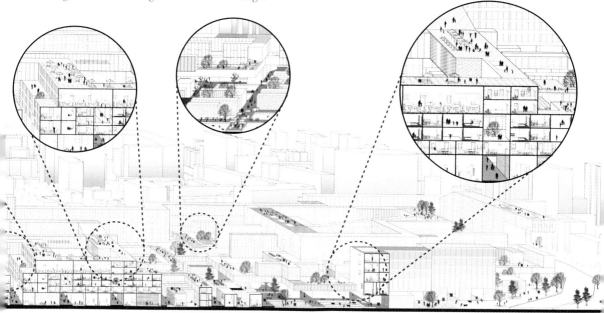

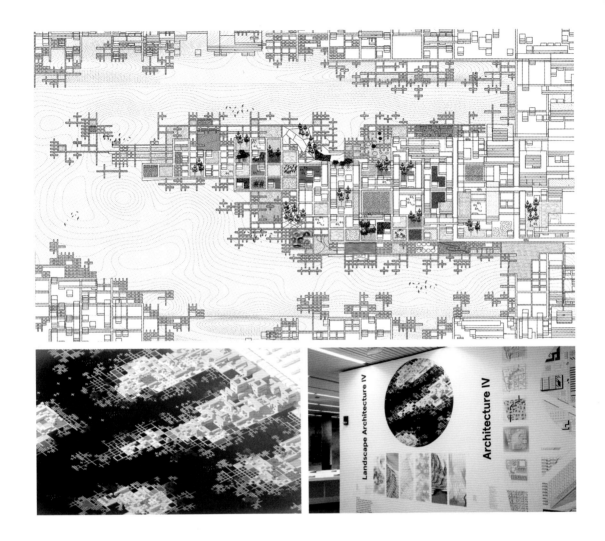

The Grid City In Transformation

We aim to design an urban framework that interrelates urbanization and ecological processes, which involves the artificial intervention and natural process.

Through the design of a flexible framework and changeable content, we aim to create an urban development model that adapts to time and nature phenomena. The urban space is deeply embedded in ecological processes, and it influences and interacts with them.

变化中的格网城市

城市发展几乎是一个强烈改造自然的行为，但如果存在一种城市，他与自然共生而且互相作用。那么城市也许有一种新的可能。研究中，形成一个将城市化和生态过程相互关联的城市框架。然而，生态过程是遵循纯粹的自然现象。这是一个人工干预与自然过程博弈的过程。

通过框架的设计和内容的变化，我们创造了一个随着时间和自然现象而变化的城市发展模式。它的空间受到自然过程的影响并与之相互作用。它不是纯生态驱动的城市形态，而是生态过程深深嵌入城市框架中的城市空间。

Rethinking Urban Renewal

The theme of this exhibition is "rethinking urban renewal," which reflects on China's urbanization model. Capital and power have transformed the city over time and space, causing daily life to withdraw from the city and removing obstacles to development. Urban renewal, guided by economic interests, has resulted in a stereotypical design. Architects play a dual role as tools and resistance in this process.

This exhibition not only questions how architecture can return to people's daily life during urban renewal, but also attempts to compare urban problems faced by different cities in China and find solutions.

再思城市复兴

展览以"再思城市复兴"为主题,这是对中国城市化模式的一种反思过程。资本和权力通过时空的转换改变着我们栖居的城市,迫使日常生活从城市中退席,从而扫除了发展的障碍。尤其在城市更新的过程中,以经济利益为导向的建设形成了千篇一律的设计。建筑师在当中扮演着双面的角色,一面可能是工具,一面可能是抵抗者。

这次展览不但是建筑师对城市更新中,建筑如何回归到人们日常生活的问题的追问,同时也是对中国各个城市所面临的城市问题的一个比对和寻找解决之道的一个尝试。

This page, above: 2017 UABB- REthinking urban renewal. This page, below: 2015 harvard gsd flux grid city. Opposite: 2015 harvard gsd flux grid city.

本页,上:2017 深双再思城市复兴。本页,下:2015 哈佛大学——变化中的格网城市。对页:2015 哈佛大学——变化中的格网城市。

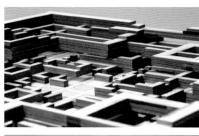

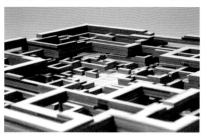

The grid itself takes advantage of the existing urban fabric and extends it as a control network. The scale and combination of the grid depend on the location of the site and its interface with nature. Through subdivision, the grid changes from being urban development-oriented to being human and nature scale-oriented.

网格利用现有的城市网络延伸,并试图将其扩展为控制城市发展的框架。它的规模和组合取决于场地的位置及其与自然的界面。通过细分的方法,网格从以城市发展为导向转变为以人为本、以自然尺度为导向。

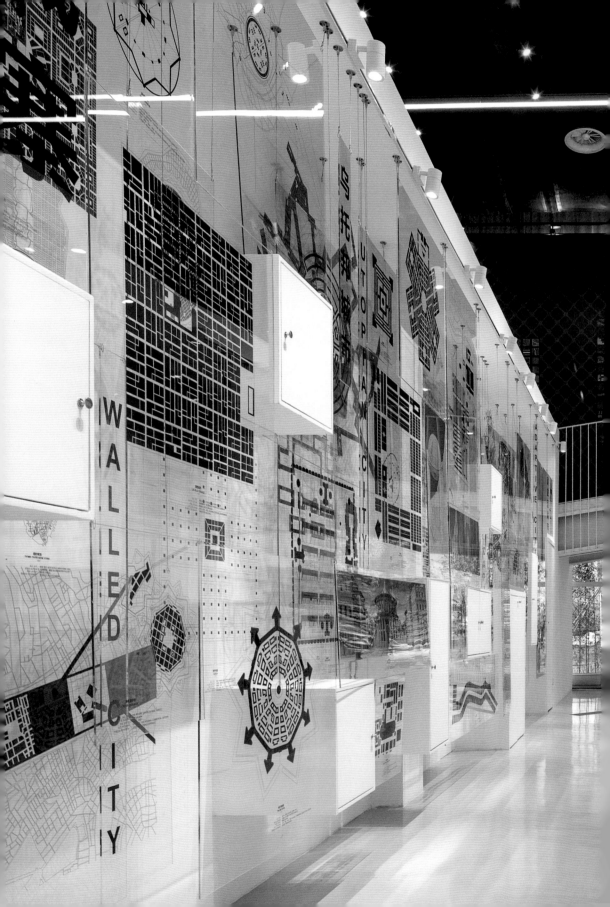

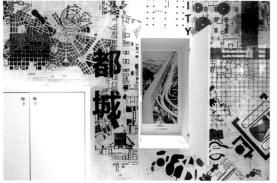 

This page, above: 2019 UABB Guangming Venue Anchor point for the public space. This page, below, Opposite: 2019 UABB Guangming curator special unit— Historical future city study.

本页，上：2019 深双光明分展场作为城市公共空间的锚点。本页，下，对页：2019 深双光明分展场策展人特别单元——历史中的未来城市研究。

2019 Study of Future Cities in the History of Shenzhen's Guangming

In response to the new adjustments to city construction, Shenzhen University's Bright Branch is once again taking on the role of space research and activation, rethinking the mode of urban development. The exhibition proposes the concept of future cities in history, examines the planning status of Guangming in the past, and explores new types of cities for the construction of a brighter future city, creating an exhibition that connects history to the future and knowledge to tools.

2019 深圳光明历史中的未来城市研究

在面对新城建设的调整时，深双光明分展场再次作为空间研究与激活的角色，重新思考城市发展的模式。展览中，对光明的城市建设提出了历史中的未来城市概念，与光明过往的规划现状，与未来新型的城市形成一个从历史到未来、从知识到工具的关联的展览。

This page, above: 2019 UABB Guangming curator special unit—Historical future city. This page, below: 2019 UABB Guangming—Syracuse Architecture Pavilion: Design | Energy | Futures.

本页，上：2019 深双活动——数字城市。本页，下：2019 深双光明分展场——雪城大学建筑学院展位：设计 | 能源 | 未来

To understand urban development better, 100 typical urban design projects from the past thousands of years of history were selected, including cities that have been realized, cities we are living in, and ideal city models that have influenced the imagination of generations of cities. Studying history is not merely a nostalgic reflection of the past, but also a retrospective search for historical experiences and future imaginations to find a new starting point. Since the medieval era, the city has evolved from a human settlement of trade and commodity exchange to a more complex composition. Urban functions have become separated, emphasizing functional division, with modern industrial cities linking work and life, and production and work becoming interlinked. The function of the city has evolved from a simple human settlement to a complex organism. Consequently, the creation, design, and construction of cities have been continuously changing throughout history. In the face of an uncertain future, perhaps history is a mirror, and historical cities have provided valuable experiences in urban construction that are worthy of constant review.

研究与展览的核心是从过去数千年的城市发展历史中选取的100个典型的城市设计,其中既包含已被实现,我们正在生活其中的城市,也包含未被实现却一直影响着世世代代城市想象的理想城市模型。研究历史并不仅仅是一种对过去的缅怀,而是以回顾的方式在历史的经验和未来的想象之间寻找一种新的触发点。从中世纪的城市开始,城市便从贸易、商品交换的人类聚居地向更复杂的构成发展。城市功能由原先强调功能分区,工作、生活、生产等截然分隔,到现代工业城市工作与生活互联,生产与工作互通的方式,迈向复合高效生态的城市发展。城市的功能从简单的人类聚居地发展成为一个异常复合的有机体。因此,城市的创造、设计和建设一直都在历史的时间线上持续更迭。面对未来的不确定,也许历史是一面镜子,历史城市是我们值得不断回顾的城市建设经验。

This page: 2017 UABB Guangming invited unit.
本页:2017深双光明分展场邀请单元——光明回溯网格城市。

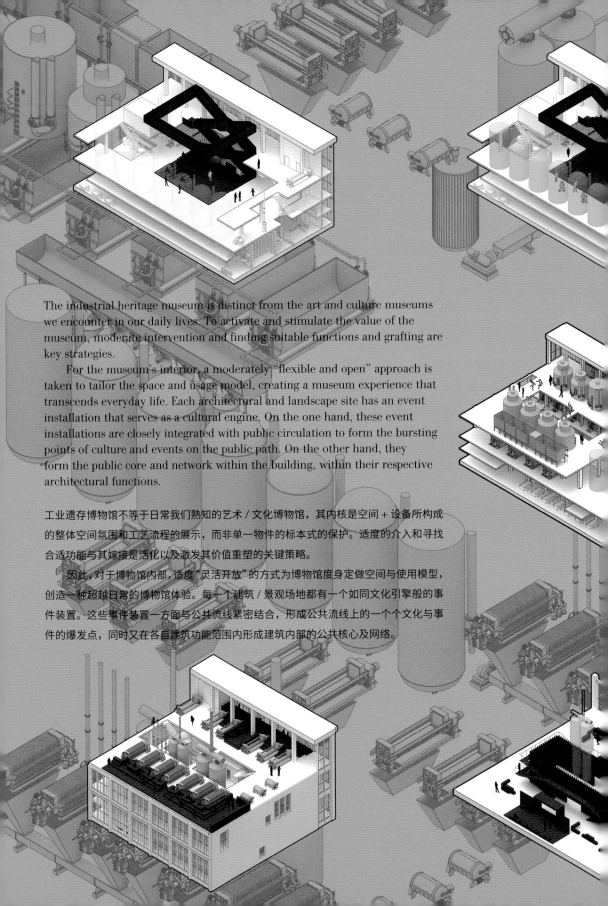

The industrial heritage museum is distinct from the art and culture museums we encounter in our daily lives. To activate and stimulate the value of the museum, moderate intervention and finding suitable functions and grafting are key strategies.

For the museum's interior, a moderately "flexible and open" approach is taken to tailor the space and usage model, creating a museum experience that transcends everyday life. Each architectural and landscape site has an event installation that serves as a cultural engine. On the one hand, these event installations are closely integrated with public circulation to form the bursting points of culture and events on the public path. On the other hand, they form the public core and network within the building, within their respective architectural functions.

工业遗存博物馆不等于日常我们熟知的艺术/文化博物馆，其内核是空间+设备所构成的整体空间氛围和工艺流程的展示，而非单一物件的标本式的保护。适度的介入和寻找合适功能与其嫁接是活化以及激发其价值重塑的关键策略。

因此，对于博物馆内部，适度"灵活开放"的方式为博物馆度身定做空间与使用模型，创造一种超越日常的博物馆体验。每一个建筑/景观场地都有一个如同文化引擎般的事件装置。这些事件装置一方面与公共流线紧密结合，形成公共流线上的一个个文化与事件的爆发点，同时又在各自建筑功能范围内形成建筑内部的公共核心及网络。

Jiangmen sugar factory renovtion and curation

Isolated sugar workshops and packaging warehouses cannot simply accommodate and meet future needs. To highlight the spirit of the place and become a carrier of new culture, they must be integrated, transformed, and incorporate into the city while providing a complete and flexible spatial model. By studying the original building layout of the Ganhua Plant, we know that the core building of Gantian Plaza consists of a pressing workshop, a sugar workshop, a packing room, and a sugar warehouse. Therefore, by restoring the spatial shape of the pressing workshop, establishing an urban public passage of the industrial museum with the sugar workshop as the main body, and integrating the original mezzanine system, we connect the public entrance of the packing room and the sugar warehouse corridor in series. In this way, we form a sugar-making space flow line in history, as well as a core public experience stream of the museum.

江门甘化厂工业遗存改造及策展方案

孤立的制糖车间和打包间仓库无法简单地容纳和承载未来的需求，只有将其整合、转化和融入城市，同时提供既完整又灵活的空间模式才能凸显其场所精神和成为新文化的载体。通过对甘化厂原建筑布局研究，甘甜广场核心的建筑由压榨车间、制糖车间、打包间及糖仓库构成。因此通过恢复压榨车间的空间形制，建立制糖车间为主体的工业博物馆城市公共通道，整合原夹层系统，串联打包间公共入口及糖仓库连廊，形成一个既是历史中的制糖空间流线，也是该博物馆核心的公共体验流线。

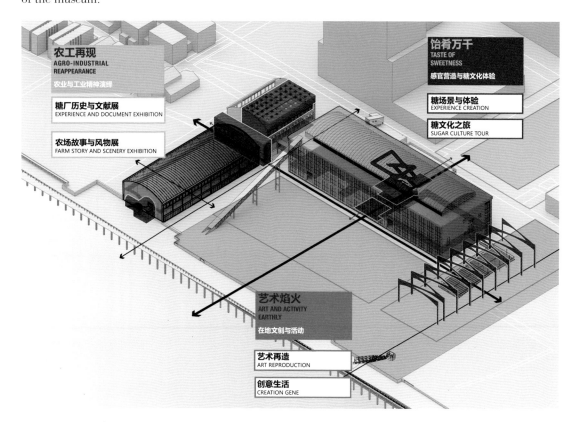

This page: Events and space curation. Opposite page: Immersive workshop space.
本页：事件与空间策动。对页：沉浸式厂房空间。

NEXTMIXING

NEXTMIXING—The Spatial Experiments on Usage
Shanghai, China 2016-Now

那行
那行——使用的空间实验
中国，上海 2016-现在

Nextmixing is a space that was separated from Atelier Archmixing, an architectural studio based in Shanghai, during the "New Year New Office" program. It is used for cultural communication and space operations. From its establishment in 2016, it was founded on a simple question: "Can such a plain space survive in the market for different uses?" Now, from Nextmixing 1.0 (2016-2019) to Nextmixing 2.0 (2019-), this space accommodates various activities, including design, illustration, acoustic art, performance, literature, food festivals, and more, with a frequency of nearly 100 events per year. Through its operation, Atelier Archmixing has not only gained an understanding of the way spatial usage works, but also deeply grasped the real meaning of spatial design and usage in Chinese contemporary urban context. Here are some representative spaces.

"那行文化（现更名为米行文化）"是在阿科米星建筑设计事务所"一年一个工作室"计划过程里分离出来的独立空间，用作文化交流与空间运营。从 2016 年成立之初，仅仅基于一个简单的问题："单纯的空间，能在市场里满足不同的使用而存活吗？"如今，经过了那行文化 1.0（2016-2019）与那行文化 2.0（2019-），每年以近百场的频率举办着设计、插画、声音、表演、文学、美食等各种活动，"那行文化"的运营实践不仅在建筑学中让阿科米星多了一重就近观察空间使用方式的维度，更直接、深入地理解中国城市环境下空间设计与使用的真实含义。以下是关于"一年一个工作室"项目想与大家分享的具有代表性的若干空间。

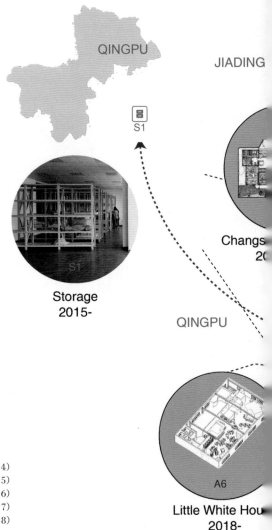

A1. Changshun Rd. Office (2009-2014)
A2. Guiping Rd. Office (2014-2015)
A3. Longcao Rd. Office (2015-2016)
A4. Hongkou Office (2016-2017)
A5. Xujiahui Office (2017-2018)
A6. Little White House (2018-)
S1. The storage in Qingpu district (2016-)
N1. NEXTMIXING: A multi-functional gallery (2016–2019)
N2. NEXTMIXING: (2019–)

A1. 长顺路工作室（2009-2014）
A2. 桂平路工作室（2014-2015）
A3. 龙漕路工作室（2015-2016）
A4. 虹口港工作室（2016-2017）
A5. 徐家汇工作室（2017-2018）
A6. 桂林路小白楼（2018-）
S1. 青浦仓库（2016-）
N1. 那行文化（2016-2019）
N2. 那行文化（2019-）

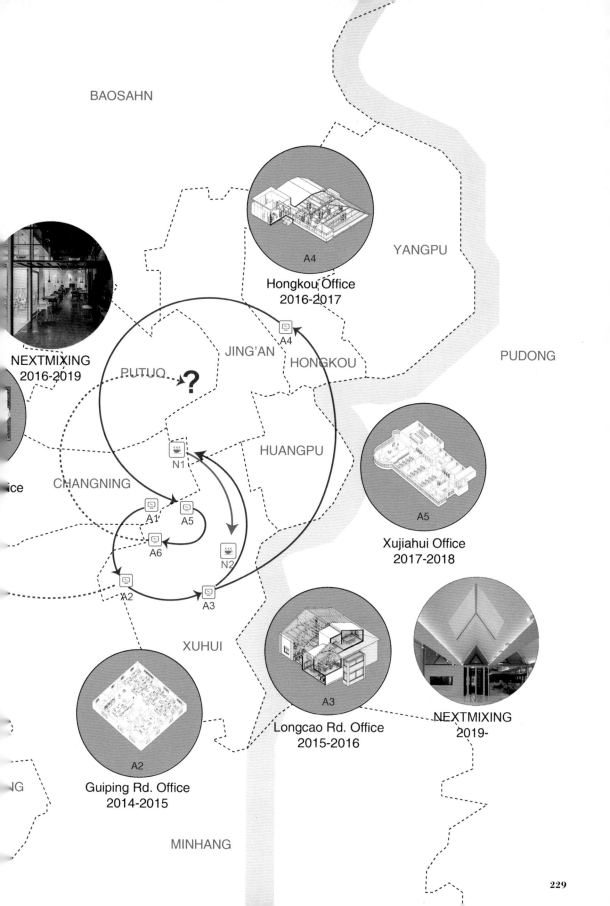

The Longcai Road Office (2015-2016)
龙漕路工作室（2015—2016）

This office was previously a factory-cum-photo studio, an urban space with visible traces of a wedding photography company. The previous tenants had divided the large building, which was nearly 6 meters high, into upper and lower areas, with many windows offering stunning views. The sounds of light rail trains filter through the office. Archmixing largely retained the spatial layout and added only a few functional areas to create a different spatial experience, intended to be independent of the style rather than constrained to it. The new layout has also influenced the way the architects work. The main architects work upstairs and can look down through control windows into the large working and conference area. At the same time, they still need to move through the space several times a day for personal communication with the employees below.

这是一个由厂房改造的摄影棚。近6米层高的厂房大空间被从事婚纱摄影的前租客分隔得高高低低、上上下下。室内很多窗口有趣地贯通着，坐在办公室里，时不时就能看见、听见紧挨的城市轻轨呼啸而过。调整设计基本保留了原来的空间格局，只在局部设计了新的功能区。这些局部的改变不求风格统一，有点各自为政，以期形成不同体验的空间效果。新的布局再次影响了我们的工作方式：主持建筑师们变得高高在上，可以透过楼上的控制窗瞭望楼下的大工作间和会议区，若要直接交流则需要经过数次上上下下的空间转换。

This page, above: Longcao Rd. Office. This page, below: A day A World exhibition after the move. Opposite: Longcao Rd. Office.

本页，上：龙漕路工作室。本页，下：搬家前的"一天世界"展览。对页：龙漕路工作室。

The Hongkou Office (2016-2017)
虹口区工作室（2016—2017）

This page, above: Hongkou Office. This page, below: Hongkou 1617 exhibition after the move. Opposite: Hongkou Office.
本页，上：虹口工作室。本页，下：搬家前的"一天世界"展览。对页：虹口工作室。

Situated on West Jiangwan Road in Hongkou District, Archmixing's fourth studio is in close proximity to several significant landmarks such as Hongkou Stadium, Hongkou Park, Longemont Commercial Complex, and the historical and cultural blocks of Hongkou. The surrounding streets and communities are lined with businesses and restaurants. The studio offers a comfortable working environment, featuring a loft that overlooks the surroundings and the nearby metro, similar to the previous studio. It's a bustling environment every day. Originally separated from the main body, a section of the studio has been connected with a small overhead gangway. A flexible desk arrangement makes it easier for staff members to communicate and create a comfortable atmosphere. In this studio, Archmixing completed its largest urban survey, "Hongkou 1617 Analytic Restoration," as well as studying the partial and fragmented changes in the surrounding urban buildings. An exhibition was held on the empty studio at the end of this cycle.

这个工作室位于虹口西江湾路，周围紧邻着虹口体育场、虹口公园、龙之梦商业综合体、虹口的历史文化街区，以及大大小小各种日常商业、餐饮丰富的街道与社区。工作室的主体是一个阁楼，可以眺望周围的环境，与前一个工作室相似，地铁三号线还在不远的视野内，每天充满了活力。工作室的另外一部分与主体原来并不连通，我们做了一条小小的架空甬道将二者联系了起来。这个工作室采用了随意分布、随时改变与组合的桌面布置，这个布置深受大家欢迎，让大家感到交流更方便，环境更轻松。在这个工作室，我们完成了历次中最大规模的城市调研——"虹口1617逆向还原"，研究了周边各种类型的城市建筑的局部与片段的改变。在结束这个周期的工作室时，我们在再次搬空的工作室里举办了一个展览。

Nextmixing (2009-Now)

那行文化（2009-现在）

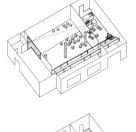

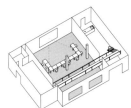

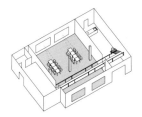

Nextmixing - Polysemous space usage pattern / 那行多义空间使用模式

"Nextmixing 1.0" (2016-2019) is located in a creative park that was once a lane factory in the historical district of Yuyuan Road. The original factory space was divided into three halls: black, white, and gray, which could be used separately or in combination. The mezzanine floor design allows for different possibilities and innovative uses of the space. After more than three years of use, it has truly demonstrated the operability of a multifaceted space. "Nextmixing 2.0" (2019-) is situated in the West Bund Art and Cultural District of Shanghai. It continues the 1.0 model's relationship and interactive use of space expansion and urban public greening.

"那行文化 1.0"（2016-2019）选址在上海愚园路历史街区一个弄堂工厂改造的创意园区，原厂房空间被分割为黑白灰三厅，空间使用上可分可合，因为设计了夹层，空间丰富，为各样活动提供不同可能和创新的使用，并在三年多的使用里，真实演绎了多义空间的可操作性。"那行文化 2.0"（2019 年至今）位于上海西岸艺术文化区，延续了 1.0 模式的关于空间扩张与城市公共绿化的关系与互动使用。

This page, left: Nextmixing 1.0, at Yuyuan Road. This page, right: Nextmixing 2.0, at WestBund art and cultural district. Opposite: <YiTiao> Consultant Video.

本页，左：那行文化 1.0, 位于愚园路上。本页，右：那行文化 2.0, 位于徐汇西岸艺术文化区。对页：《一条》顾问视频。

As a cross-border experimental platform involved in space operations, "Nextmixing" continuously explores new communities, scenarios, and experiences. This is reflected in two main ways:

Firstly, since the creation of Yitiao Video, TANG Yu, the director of Nextmixing, has been an advisor to its architecture program, greatly contributing to the recognition of outstanding works of contemporary Chinese architecture by the general public. The two video series curated by TANG Yu, *Homage to Classic Architecture* and *The New Wave of Chinese Architecture*, have promoted and popularized the best local architects and their works in China since modern times. These videos offer a lifestyle-oriented and popular way for ordinary people to get to know and understand architecture, showcasing nearly 200 architectural stories to hundreds of millions of viewers.

Secondly, Nextmixing has full-link media

"那行文化"作为含有空间运营的跨界实验平台，不断地探索新社群、新场景和新体验。其实践在以下两个方面有完整的体现：

第一，那行文化主理人唐煜自一条建筑视频创立以来，担任其建筑栏目顾问，极大地推动了中国当代建筑优秀作品被大众认知。唐煜策划的"致敬经典建筑"和"中国建筑新浪潮"两个系列视频节目，为观众推广并普及了中国近现代以来的优秀本土建筑师及其作品，视频以生活方式为导向，用大众的方式引领普通人认识、了解建筑，向上亿的观众讲述了近200个建筑故事。

第二，那行文化拥有全链路媒体策划能力。自第一届参与到上海城市空间艺术季（SUSAS，创立于2017年，每两年一届，以"文化兴市，艺术建城"为理念，通过艺术季活动达到改善生活空间品质、提升城市魅力为目标）以来，通过每届的合作逐步完成媒体宣传策划、分展场展览策划与执行、专项特色活动的组织策划执行，乃至产品设计等方方面面的工作。

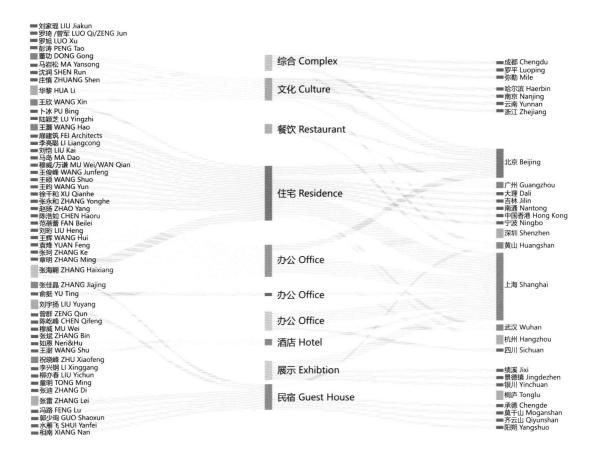

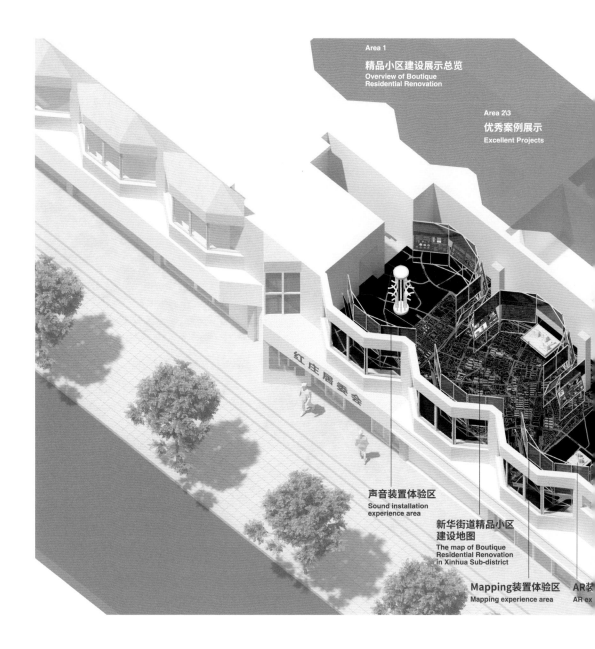

planning capabilities. Since participating in the first Shanghai Urban Space Art Season (SUSAS) in 2017, a biennial event with the concept of "culture to promote the city, art to build the city," and the goal of improving the quality of living space and enhancing the charm of the city through art season activities, we have cooperated and completed media publicity planning, exhibition planning and execution, special events organization, planning and execution, and even product design.

Invited by the 2021 SUSAS, Nextmixing served as a joint curator and exhibition designer, transforming the community office into an exhibition and forum space that can quickly restore its functions—the "Suitable for Living·LOHAS" experience gallery. The exhibition's theme is "The Action of Suitable for Living" and "My Community," and through the use of self-illuminated light-weight display frames, integral floor stickers, AR, and other interactive devices, it showcases "livable" community work methods to the public. It provides an in-depth analysis of excellent cases of boutique

Area 4
优秀案例展示
(兼做沙龙分享区)
Excellent Projects
(Salon space)

应2021上海城市空间艺术季邀请，那行文化作为联合策展人、展览设计师，将社区办公室改造为可快速复原功能的展览与论坛空间——《安居·乐活家》体验馆。展览以"安居行动"和"我的小区"为主题，通过自照明的轻质展架、整体地贴、AR等互动装置，向广大市民展示社区"宜居"的工作方法、深度解析精品小区改造的优秀案例，传递沉浸式的城市空间艺术展览体验。

在这些全链路策划实践中，那行文化越来越认为自身角色的意义是能够承担政府与大众的沟通桥梁，将上层建设、时代精神转译为与广大市民日常生活息息相关的内容，将专业知识化作通识教育，创立共建共商共创的多元平台。

community transformation and delivers an immersive urban space art exhibition experience.

In this full chain of planning practice, Nextmixing increasingly sees the significance of its role as a bridge between the government and the public. We translate the construction of the upper echelons and the spirit of the times into content that closely relates to the daily lives of the general public, turning professional knowledge into general history education and creating a multifaceted platform for joint construction and co-creation.

Opposite: Exhibition Axonometric Drawing. This page: A variety of interactive installations on display. The "Living LOHAS Exhibition" experience pavilion, the sound installation "Resonance" in collaboration with sound artist YIN Yi, the AR installation "Participatory Design" curated by technology company (Bitsreal Technology) - an art installation that uses mapping technology to inspire the viewer to participate in the construction of a community living environment; and "My Neighbourhood is Renewing" - an augmented display using AR technology to help the public visualise the construction of a quality neighbourhood.

对页：展览轴测图。本页：展览上的多种互动装置。那行与声音艺术家殷漪合作声音装置《共鸣》、科技公司（彼真科技）策划AR装置《参与式设计》——利用mapping技术，激发观者共同参与社区居住环境建设的艺术装置；《我的小区在更新》——利用AR技术，以增强显示的手法帮助公众直观地了解精品小区建设。

237

Archild
Archild
Shenzhen, China, 2013-Now

童筑文化
童筑文化
中国,深圳 2013-现在

Archild's goal is not to train future architects. In China, there is already an oversupply of architects, and using architecture to meet people's needs is a too ancient approach. However, studying architecture can help us understand the relationship between people and the surrounding environment, and can inspire our imagination and creativity. The method of studying architecture is also unique. It is not based on the mind, but on the perception and labor of the body. In other words, Archild emphasizes learning through the body, and learning is also a process of discovering and developing our own bodies. Therefore, architecture can be a part of everyone's quality education.

pp. 238-239: The children are building a spatial framework device using PVC water pipes.
This page, above: Children use bamboo weaving to build a shelter that can accommodate their own bodies. This page, left below: Children transformed into little planners to participate in an urban planning design competition. This page, right below: The children designed and created a three-dimensional city installation together, and arranged the lighting.

第 196-197 页：童筑的孩子们在用水管搭建一个空间框架装置。
本页，上：童筑的孩子们用竹编织搭建一个能容纳自己身体的庇护所。本页，左下：童筑的孩子们化身小小规划师在进行一场城市规划设计竞赛。本页，右下：童筑的孩子们一起设计制作了一个立体城市的装置，并布置了灯光。

筑的目的不是培养未来的建筑师，现在中国的建筑师已经过剩了，而且用建筑来满足人们的需要是个太古老的方式。但是学建筑可以帮助我们理解人与身边环境的关系，可以激发我们的想象力、创造力。同时学习建筑的方法也是特别的，不是靠头脑，而是靠身体的感知和劳动，也就是说我们强调用身体去学习，在学习中也是我们发掘和发展自己身体的过程。所以建筑学可以是每个人的素质教育的一部分。

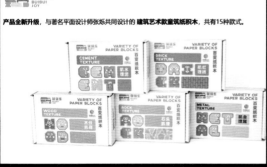

This page, above, right below: The VI visual system of Archild.. This page, left below: A children's building toy ,the Paper Blocks, that was designed and produced by Archild.

本页，上，右下：童筑文化的VI视觉系统。本页，左下：童筑文化设计制作的一款儿童建筑玩具：百变纸积木。

Zidao Culture
Shituzi Architecture Channel
China 2013-2021

子道文化
使徒子建筑频道
中国 2013-2021

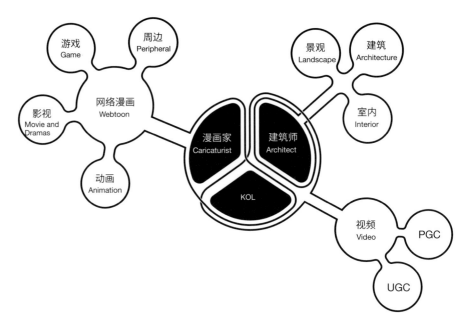

As an architect, QIN Qingxia has created an online video series that focuses on design aesthetics, collaborating with some of the most prestigious firms in the industry. During the 2022 Beijing Winter Olympics, QIN worked with Bilibili Encyclopedia and the Winter Olympics Committee to produce videos on design and architecture in the games. The total number of views on Weibo and Bilibili has exceeded 10 million, and the videos have been at the top of the trending list several times, helping to promote the Winter Olympics and winter sports in the media. This video series has frequently ranked at the top of Weibo's trending list and attracted widespread attention to the field in China.

使徒子凭借自身建筑师身份，开发全新的视频产品，聚焦设计美学科普，并和行业里的很多大咖进行了合作。在冬奥会期间，使徒子联合B站百科和冬奥奥组委合作产出的科普视频，在微博、B站总播放量过千万，多次登上热搜榜，助力北京冬奥会和冰雪运动。自开展设计美学科普视频以来，频频登上微博高位热搜，获得了广泛的关注。

This page: Architect QIN Qingxia interviews guests as a media worker. Opposite page: Architectural knowledge popular content with relatively high click volume on the video media platform.

本页：建筑师覃清硖以媒体人的身份采访行业嘉宾。对页：媒体视频平台上点击量比较高的建筑知识普及内容。

朝鲜建筑怎么样?【使徒子】
▷ 25.5万　4-29

老头环的建筑设计依据是什么?【使徒子】
▷ 5.8万　4-22

青瓦台的风水好不好?【使徒子】
▷ 39.5万　4-15

什么样的近代艺术大佬会画美少女还捏手办?【使徒子】
▷ 8.3万　4-12

最离谱的现代主义大师是谁?【使徒子】
▷ 4.8万　4-1

建筑学生就业怎么样?【使徒子】
▷ 18.9万　3-29

普利兹克奖首位非洲获奖者是什么来头?【使徒子】
▷ 20万　3-25

现实中也有纪念碑谷!建筑梦想家的作品【使徒子】
▷ 6.2万　3-18

没有他们就没有中国的建筑专业!梁思成与林徽因的传奇奋斗
▷ 5.6万　3-8

德式美学是怎么来的?【使徒子】
▷ 15.7万　3-4

建筑界第一神仙眷侣是谁?【使徒子】
▷ 3.8万　2-25

印度的神奇建筑长啥样?【使徒子】
▷ 120.4万　2-18

勃列日涅夫楼长什么样的?【使徒子】
▷ 37.5万　2-12

如何用建筑搞教育?【使徒子】
▷ 4.2万　2-8

是鬼才建筑师?还是阴间建筑师?【使徒子】
▷ 36.2万　2-5

工业遗址上的滑雪大跳台
▷ 31.6万　2-2

冬奥选手住区有多高级?【对谈设计师】
▷ 64.5万　1-28

08奥运老将新生,水立方华丽变身冰立方【对话设计师】
▷ 25.6万　1-26

冬奥速滑馆冰丝带,竟藏着这么多秘密!【对话设计师】
▷ 25.3万　1-24

这届冬奥会的设计也太酷了!【使徒子】
▷ 15.6万　1-21

把房子当机器造是种什么体验?【使徒子】
▷ 6.5万　1-7

辣眼睛!去年最丑建筑都有谁?【使徒子】
▷ 178.9万　1-2

古建应该怎么看?【使徒子】
▷ 14.6万　2021-12-31

现代主义最后的大师是谁?【使徒子】
▷ 6.2万　2021-12-24

Interview: Aric CHEN
After Architecture

访谈：陈伯康

建筑之后

WANG Fei: Why after training in architecture professionally at Berkeley you went to graduate school in museum study and became a professional curator?

Aric Chen: I grew up in Chicago, and architecture is really in the blood there. But though I was first seduced by all the Mies van der Rohe and Frank Lloyd Wright buildings I grew up around, architecture was for me a first lens for looking at the wider world, the built and constructed environment. "Constructed" is in both the literal and anthropological senses. In fact, at Berkeley, I soon added anthropology as a degree, and did that and architecture at the same time.

How I wound up becoming a curator is a bit more convoluted. I did my masters in a museum, at the Cooper-Hewitt in New York, in design history—not curatorial studies. Curatorial studies wasn't such a thing back then, and to be honest, it wasn't an interest in curating that brought me to Cooper-Hewitt. I'd studied space and the way we operate as humans and cultures within and around space, but I felt I was sort of missing the object; hence, design history. However, I wasn't conscious of how architecture, anthropology and design were going to come together for me. These were just components of my broader curiosity about the world we inhabit.

And it was through that curiosity that I sort of accidentally fell into journalism and writing in New York, which is how I got into curation, and that's a whole other story. If you look at many of the architecture and design curators and museum directors out there, a lot of us came from journalism.

WANG Fei: As a curator, you have curated many exhibitions on architecture, urbanism, design and art, what do you think the relationship between curation and all these disciplines and industries?

Aric Chen: To go back to curation and journalism, to me, they are really of a piece. They're fundamentally both about seeing and observing as much as you can, situating it in its context, filtering and editing what you see, gleaning insights, drawing conclusions, and building it all into narratives for an audience.

But on the question of disciplines, there are of course specific disciplinary trajectories that have their own structures and histories that have accrued over time through the development of institutions and other frameworks, largely in the west. These have come to define what we now think of as disciplines, but we need to remember them for being the somewhat limited constructs that they are. So going back to my sort of subconsciously interdisciplinary educational background, these disciplines were perhaps not so much components of a bigger picture, but again, lenses or starting points for looking at even broader landscapes.

So you can look at the world through urbanism; you can look at the world through art; you can look at the world through curation. But I think it's important to always understand that these lenses are not absolute truths. Like urbanism as

王飞：什么原因促使您经受了严格传统的建筑学设计学教育，而去读了博物馆研究的研究生，做了职业策展人？

陈伯康：我在芝加哥长大，建筑在那里有很深的渊源。尽管小时候生活的环境使我对密斯·凡德罗以及弗兰克·赖特设计的建筑着迷，但建筑对于我来说是审视更广阔世界、建筑以及建造环境的第一个镜头。"构成"指的是文学和人类学的双重意义。事实上，考入加州大学伯克利分校之后，我同时在修人类学和建筑学的双学位。

关于后来我成为策展人的过程是比较复杂与曲折的。我在纽约库伯·休伊特博物馆攻读硕士学位，专业是设计历史，而不是策展方向。当时，策展本身还不是一个专业，老实说，我也不是因为策展而去的库伯·休伊特。我之前已经学习空间以及我们作为人类和文化在空间内外的运作方式。但与此同时，我感觉还缺少一个目标，因此，我决定学习设计历史。但在当时，我并不确定建筑、人类学以及设计应该如何融合在一起。这些只是我对于我们的生活环境产生的无尽好奇心的一些片段。

正是通过这样的好奇心，我在纽约意外地接触了新闻学和写作，这也是我最终了解策展的原因和过程。你看许多建筑、设计策展人和博物馆馆长，他们中的很多人都来自新闻业。

王飞：作为策展人，您也做了大量的展览与建筑、城市规划、设计、艺术相关的工作，您觉得策展与这些行业的关系是如何的？

陈伯康：回到策展以及新闻学，对我来说，他们是完全一致的。从根本上来说，它们都是尽可能多地观察，并将其置于背景中，同时过滤以及整理所见所闻，得出结论，最后通过叙事故事讲给观众们。

但在学科问题上当然有特定的学科发展轨迹，每一个学科有自己的结构和历史，随着时间的推移，通过制度和其他框架的发展而积累起来的，并且大多在西方。它们定义了我们认知的这个学科，但我们也需要记住它们自身的局限性。所以回到我那种潜意识的跨学科教育背景，这些学科可能无法组成更大的愿景，但却是成为观察更广阔风景的视角或起点。

所以你可以通过都市主义看世界；你可以通过艺术看世界；你可以通过策展看世界。但我认为重要的是始终知道，通过这些镜头看到的并不是绝对真理。就像都市主义学科一样，它并不是真理，而是一种建构的方式与方法。正如我们所知，这些学科长期以来一直在尝试扩展自己的界限，并与其他学科跨界交流。我想策展也是一样，这些意识都比我们通过教育获得的认知要开放得多。

王飞：您受聘于同济大学设计与创意学院，开设了策展的课程及研究生方向，是如何与传统的设计课程结合的？或者是打破了传统的模式？在这几年的教学当中，设定的目标是什么？是否达到了？

陈伯康：我是否实现了雄心壮志不是我说了算，但我可以肯定，雄心是有的！那就是重新思考如何不仅仅只是理解策展，更是在当代中国的背景下以全球视野来理解

245

a discipline is not a truth, but a construct. As we know, these disciplines themselves have for a long time been expanding their own boundaries and working across their boundaries with other disciplines. I would just put curation within that. It's all much more fluid than we've been historically trained to believe.

WANG Fei: You were hired at College of Design & Innovation at Tongji University, Shanghai, and you started the courses in curatorial both in bachelor and master levels. How do you work with the traditional curriculum in design school? Did you break the conventional model? What was the goal or aspiration in those years' teaching? Have you achieved the original ambition?

Aric Chen: Whether I achieved the ambition is not up to me to say, but I can confirm that the ambition was there! And that was to rethink how one might approach not curation necessarily, but curatorial subjects in the context of contemporary China with still a global perspective.

I was fortunate to have been invited to teach at Tongji. It's an incredibly dynamic place, but more to the point, a very open one. The dean, LOU Yongqi, had a progressive and open mindset that allowed a lot of room and freedom for faculty, staff and students. I was asked to teach a design history course and a research studio and had the room to try different approaches to these subjects and formats.

For design history, this was kind of a continuation of the work that we had started when I was at M+, the museum in Hong Kong that opened last year, where from 2012-2018 I was lead curator for design and architecture responsible for overseeing the formation of the museum's design and architecture collections, programs and so on. What we were trying to do there was to recenter design histories through the collection, the objects, the materials that we acquired.

On the one hand, this was about making more visible some of the lesser-known narratives of design and architecture in Asia, narratives that are not as well-known because institutions—whether in Asia or the West—have not been telling them, at least not in very consistent and public ways. To a large extent, this was the raison d'etre of M+, to build a collection that could be the backbone for developing and amplifying these narratives, whether they be the history of contemporary Chinese art, post-World War II Japanese graphic design, post-Independence architecture in Southeast Asia or whatnot.

On the other hand, it was about revisiting well-known global narratives from the vantage point of our region. For example, if you look at postmodern architecture and design from the vantage point of Hong Kong or Japan, they become something very different than if you look at them from the traditional postmodernism discourse centers of Italy, UK, and the US. So, we wanted to reexamine some of these ideas from situated Asian perspectives and see what new understandings we could draw from that.

With the design history course at Tongji, we were trying to work with students to look at design history from a Chinese-centric perspective. Instead of teaching them about the Bauhaus, which was indeed influential in China, in an orthodox way—that is, the ideas of Weimar and Dessau spreading through Gropius at GSD, Mies and Moholy-Nagy in

Exhibition View, *Shifting Objectives: Design from the M+ Collection*. Courtesy of M+, Hong Kong

展览场景，"形流意动：M+ 设计藏品"，图片由 M+ 提供

M+, Hong Kong. © Virgile Simon Bertrand. Courtesy of Herzog & de Meuron and M+
香港 M+ 大楼 © Virgile Simon Bertrand，图片由 Herzog & de Meuron 和 M+ 提供

策展主题。

我有幸受邀在同济任教。这是一个令人难以置信的充满活力的地方，但更重要的是，一个非常开放的地方。娄永琪院长具有进取和开放的心态，为教职员工和学生提供了很大的空间以及自由。我受邀教授设计史课程和研究工作室，并有机会尝试不同的方法来处理这些主题和格式。

就设计史而言，这是我们对去年开放的香港 M+ 博物馆工作的延续，从 2012 年到 2018 年，我担任设计和建筑的首席策展人，负责监督博物馆的设计和建筑藏品、项目等。在那里，我们试图通过获得的收藏、物件和材料来重新审视设计历史。

一方面，这是为了让亚洲一些鲜为人知的设计和建筑叙事更加引人注目，因为无论是亚洲还是西方的机构都没有以非常一致和公开的方式告诉大众，所以这些叙事并不为人所知。所以在很大程度上，这就是 M+ 存在的理由，来建立一个可以成为发展和放大这些叙事的展览系列，无论这些叙事是中国当代艺术史、二战后日本平面设计、东南亚独立后的建筑历史还是其他类似项目的中坚力量。

另一方面，它是关于站在我们地区的有利位置重新审视一些知名的全球叙事案例。例如，如果你从中国香港或日本的角度来看后现代建筑和设计，它们与你从传统后现代主义中心的意大利、英国和美国审视角度出发会非常的不同。因此，我们想从亚洲的角度重新审视其中的一些想法，看看我们可以从中得出什么新的理解。

在同济大学的设计史课程中，我们试图与学生一起以中国为中心的角度看待设计史。我们没有以正统的方式向他们讲授在中国确实有影响力的包豪斯，所谓正统的方式，是指魏玛和德绍的思想通过哈佛大学设计学院的格罗皮乌斯、芝加哥的密斯和莫霍利 - 纳吉、乌尔姆的马克斯比尔传播的方式，而是将包豪斯放在背景当中，成为当时思想网络中的一个不那么中心的节点。他与中国的联系主要是通过东方集团、苏联和日本。

Chicago, Max Bill at Ulm—we placed the Bauhaus in the background as a slightly less central node in the network of ideas that were operating at the time, and that interacted with China mostly through the Eastern Bloc, Soviet Union and Japan.

And then we challenged students to look at different objects across time spans, media, and disciplines, to draw unexpected connections. We asked them to not just look at graphic design, for example, but to create seemingly unrelated pairings. We wanted to encourage them to think beyond disciplinary boundaries and see less obvious relationships.

Meanwhile, the research studio was about developing new curatorial strategies to respond to contemporary Chinese contexts. Over my 13 years in China, there was a lot of talk and work around building a curatorial infrastructure, training curators, and so on and so forth. I think a lot of times this wound up being about simply importing western notions of curation and "best practices" to China. There was value to that, but it seemed insufficient and sometimes misguided because those western modes of curation evolved from a totally different set of institutional and historical circumstances. Therefore, we wanted to see what other forms of curation might emerge from the particularities of China, and one of our studios was looking at the convergence of art, shopping, and entertainment. Of course, Rem Koolhaas and his students at Harvard were examining this 20 years ago and helped inspire this studio, but we wanted to look at how the rise of e-commerce, social media, wanghong culture and other factors specific to China could shape, and be shaped by, curatorial practices through a critical examination of the "art mall" phenomenon in China: the very splashy emergence of "art malls" like K11, SKPS in Beijing, TX Huaihai in Shanghai. The students were tasked with breaking apart and revisiting notions of curation within these hybrid retail-entertainment-cultural complexes, to examine the impact of these spaces on conceptions of art and how it relates to audiences in both problematic and interesting ways—spatially, curatorially, systemically and so on.

Putting art in a shopping mall is a longstanding practice, beginning with the first malls in the US in the post-World War II period. But China's art malls take this to another level, with the very term—"art mall"—representing art's almost total commodification and absorption into systems of consumption, in a country where a relatively weak institutional infrastructure yields enormous cultural production power to brands and real estate developers, who hold enormous sway over what gets seen and how it is seen. This can blur the line between art and visual merchandising, which poses questions about the social and cultural value and purpose of the art that's shown. We wanted to look into this while acknowledging that the marriage of art and commerce was not going away and in fact had culturally productive potentials. It was not about fighting this phenomenon but seeing how we could work with it.

The other studio that I also really loved was called "Reverse Curating." This was a response to the massive boom in museum buildings in China over the past 15 years or so, whereby thousands of new museum buildings have been constructed, largely driven by political, bureaucratic, economic, and even architectural agendas—and perhaps less so by cultural ones.

Let me explain. Very generally speaking, cultural policy mandates local governments to build new cultural infrastructure, and often the easiest way of showing you're doing that is by building a museum. However, local governments, whose revenue disproportionately relies on land sales, don't have funding for this. So to kill two birds with one stone, what has often happened is that when a local government sells land to developers, part of the deal is that the developer will build a museum, which the latter sees as having the added benefit of enhancing the value of the residential, commercial, or office units that are of course their main objective. The local officials get what they want, a cultural building. Meanwhile, the developers get their land alongside a museum that theoretically adds value to it. It seems like a win-win result, except none of these agendas are content driven or

然后，我们要求学生以跨时间、跨媒体和跨学科的方式观察不同的对象，从而得出意想不到的联系。例如，我们要求他们不仅仅关注平面设计，还要建立看似无关的匹配关系。我们想鼓励他们跨越学科界限去思考，看到不太明显的关系。

与此同时，我们的研究工作室正在开发新的策展策略以应对当代中国的环境。我在中国的13年里，有很多关于建立策展的基础设施、培训策展人等等的讨论和工作。我认为很多时候这只是简单地将西方的策展理念和"最佳实践"引入中国。这很有价值，但它似乎不够充分，有时甚至被误导，因为这些西方的策展模式是从完全不同的制度和历史环境中演变而来的。因此，我们想看看中国自身的特殊性会产生什么样形式的策展。我们的一个工作室正在研究艺术、购物和娱乐的融合。当然，雷姆·库哈斯和他在哈佛的学生20年前就在研究这一点，并帮助启发了这个工作室，但我们想通过对中国"艺术商城"现象，如K11、北京SKPS、上海TX淮海等的批判性审视，来研究电子商务、社交媒体、网红文化以及其他中国当代独特文化现象如何塑造策展实践。学生们的任务是打破和重新审视这些混合零售-娱乐-文化综合体中的策展概念，以审视这些空间对艺术概念的影响，以及它如何系统性地在建筑空间、策展方面呈现问题并以有趣的方式与观众互动。

将艺术品放在购物中心是一种长期的做法，这开始于二战后美国的第一批购物中心，但中国的艺术商城将这一点提升到了另一个层次，"艺术商城"这个词代表艺术几乎完全商品化并进入消费系统中，在这个国家里，相对薄弱的基础设施建设为品牌和房地产开发商提供了巨大的文化生产动力。房地产开发商对所看到的东西和如何看到它拥有重要的影响力，这可能会模糊艺术和视觉营销之间的界限，从而对所展示的艺术的社会和文化价值提出质疑。我们想研究这一点，同时承认艺术和商业的结合不仅没有消失，而且实际上还具有文化生产潜力。这不是要与这种现象作斗争，而是要看看我们如何去与之共存。

我非常喜欢的另一个工作室叫做"反向策展"。这是对过去15年左右中国博物馆建筑大规模涌现的回应，在此期间，数千座新的博物馆建筑的在很大程度上受到政治、官方、经济甚至建筑议程的推动——但却很少是因为文化而出现。

我来解释一下这个现象，总的来说，政策要求地方政府建设新的文化基础设施，而展示执行这种政策要求最简单的方法就是建造文化博物馆。然而，依赖于出售土地的地方政府收入不成比例，因此并没有能力为此提供资金。所以一石二鸟，经常发生的事情是，当地政府在把土地卖给开发商的时候，一部分的交易条件是要求开发商要建一个博物馆，而后者也认为这有利于提升这块土地的价值。当然，他们的主要目标依然是住宅、商业或办公楼。当地官员得到了他们想要的东西，一座文化建筑。

Abover: Het Podium © Ossip van Duivenbode. Courtesy of Het Nieuwe Instituut. Below: Expanding Narratives 2019 © Shanghai Glass Museum.

上：屋顶制作 © Ossip van Duivenbode，图片由 Het Nieuwe Instituut 提供。下：扩展叙事，2019 © 上海玻璃博物馆。

curation driven, and so the actual cultural function of the museum slips through the cracks. Hence, we have a surplus of empty and underutilized museum buildings in China, many of which are difficult to use because they were designed simply as architectural icons with spaces that aren't fit-for-purpose because, from the outset, there was no clearly-defined purpose.

Instead of just complaining about this, can we acknowledge that these buildings are not going away and find ways to curatorially reverse-engineer them? In an ideal world, you'd start with a curatorial function for a museum and then design the building for it. But if we're starting with a building that is empty and perhaps also hard to work with because the architect was not designing for any specific program, can we look at these buildings as pre-existing conditions that can serve not as endpoints but as starting points? Can we begin with this pre-existing condition—the building, its social and urbanistic context, even its funding model—and work backwards in order to give it a social and cultural purpose?

WANG Fei: Several renowned American architecture schools have set up master programs in curatorial, e.g., Harvard University GSD's MDes Art, Architecture and Public Domain and Columbia University GSAPP's MS Critical, Curatorial, and Conceptual Practice in Architecture. What do you think of this trend?

Aric Chen: I think the emergence of these courses, also like the ones at the RCA, Goldsmiths and Kingston, have been great for bringing a lot of focus and reflection to curation and offering critical methodologies and approaches that expand its possibilities. Some great people doing amazing work have come out of these programs.

But while it's for sure critical to develop new forms of curation and curatorial knowledge, I would argue that we need to make sure that curating remains a means to an end rather than an end unto itself, that the important critical reflection we engage in doesn't stop at that. Otherwise, it becomes too much navel-gazing.

WANG Fei: If there is anything you can make to change the current architecture education, what is it?

Aric Chen: I think a lot of changes are already happening. But in general, I could argue that architecture and architectural education might benefit from being less insular, that we can further develop our cultures of building discourses and bodies of thinking within architecture for sure, but then also think and work more collaboratively with other fields of expertise, both academic, professional and non-professional. This requires a degree of humility. In architecture, we are very good at speaking with each other and amongst ourselves, but usually not so good at speaking with others—and then we wonder why no one is listening to what we're saying. We've got to break out of our shells.

This is one of the things we're placing some focus on at Het Nieuwe Instituut, as a cultural institution working towards not just a multivocal state but an inter-vocal one, whether that means bringing together designers, policymakers, activists, and entrepreneurs to tackle an urban question or assembling diversities of human and non-human voices and knowledges in bringing about a more ecologically regenerative future. It's not just about exploring ideas but enacting them—and seeing a cultural institution as being not just as a place of research, debate and presentation, but a testing ground where propositions can be tried out in something resembling a real-world scenario.

与此同时，开发商得到了土地以及一个理论上可以增加价值的博物馆。这似乎是一个双赢的结果，但问题是这些决策全都不是内容驱动或策展驱动的，因此博物馆的实际文化功能就很难得到保证。所以在中国有很多空置和未充分利用的博物馆建筑，其中许多难以被使用的原因是因为它们只是被设计为一个个没有实际用途的建筑符号，从一开始就没有明确的目的。

与其抱怨这一点，我们是否可以承认这些建筑不会消失，并找到针对它们的逆向策展的方法？在一个理想的世界里，你会从博物馆的策展功能开始，然后为此功能设计建筑。但是，如果我们从一座空荡荡的建筑开始，而且可能因为建筑师设计时没有任何特定功能而难以使用，可我们是否可以将这些建筑视为预先存在的条件，不作为终点，而是作为起点？我们是否可以从这种预先存在的条件——建筑，它的社会和城市环境，甚至它的投资模式——开始，然后逆向思考，从而赋予它一个特色的社会和文化目的？

王飞：有些美国的顶级建筑学院十几年前也创立了策展与艺术的方向，比如哈佛大学和哥伦比亚大学，您如何看待这种现象？

陈伯康：我认为这些课程的出现，就像英国皇家艺术学院、伦敦大学金史密斯学院和金斯顿大学的课程一样，会很好地为策展带来很多关注和反思，并提供扩展其可能性的关键方式和方法。从这些计划中会出现一些出色的人物。

但是，虽然研究新的策展形式和钻研策展知识肯定是至关重要的，但我认为我们需要确保策展仍然只是达到目的的一种手段，而不是目的本身。我们参与的重要批判性反思不止于此，否则，就有点捡了芝麻丢了西瓜的意思了。

王飞：如果让你对现今的建筑教育做些改变，有何建议？

陈伯康：我认为很多变化已经在发生。但总的来说，我认为建筑行业和建筑学教育可能会受益于彼此较为紧密的关系，我们可以进一步发展我们在建筑中建立的思想体系，以及相关的建筑文化，也可以与其他专业领域进行更多的思考和合作，这既包括学术的，也包括职业的和非职业的。这需要一定程度的谦逊。在建筑中，我们非常擅长与行业内的人交谈，但通常不太擅长与这个专业以外的人交谈，然后我们不理解为什么社会上没有人关注我们这个行业，所以我们必须破壳而出，必须破圈。

这是我们在荷兰新学院重点关注的事情之一，作为一个文化机构，它不仅致力于建立一个多元声音平台，而且是一个多语境平台，这不仅意味着将设计师、政策制定者、活动家和企业家聚集在一起解决城市问题，也意味着汇集人类、非人类声音以及知识的多样性，以实现更具生态再生性的未来。这不仅仅是探索想法，更是将其付诸实践，并将文化机构视为一个研究、辩论和展示的场所，更是一个可以在类似现实世界的场景中进行实验的试验场。

Aric Chen is General and Artistic Director of Het Nieuwe Instituut, the Dutch national institute and museum for architecture, design and digital culture in Rotterdam. Chen previously served as Professor and founding Director of the Curatorial Lab at the College of Design & Innovation at Tongji University in Shanghai; Curatorial Director of the Design Miami fairs in Miami Beach and Basel; Creative Director of Beijing Design Week; and Lead Curator for Design and Architecture at M+, Hong Kong, where he oversaw the formation of that new museum's design and architecture collection and program.

陈伯康现任荷兰新建筑协会馆长，该机构是位于鹿特丹的以建筑、设计、电子文化为主体的新型博物馆。在此之前，他在中国生活的13年期间，曾担任上海同济大学设计创意学院策展实验室主任、同济大学实践型教授、设计迈阿密策展总监（巴塞尔、迈阿密）、北京设计周创意总监和 M+ 视觉文化博物馆设计与建筑总策展人（香港）。此外，陈伯康还在国际上策划了数十个博物馆展览和其他项目，在许多委员会和评委会任职。他曾担任深港城市\建筑双城双年展、伦敦设计双年展、库珀-休伊特设计三年展（纽约）和光州设计双年展的顾问。他是《巴西现代》（Monacelli, 2016）的作者，并经常为《纽约时报》、《Wallpaper*》、《建筑实录》和其他出版物撰稿。陈伯康曾在加州大学伯克利分校获得建筑学和人类学学士学位，并在纽约的帕森斯设计学院／库珀-休伊特获得设计史硕士学位。

Profile

设计师简介

WANG Fei is an accomplished architect, educator, critic, entrepreneur and curator. Currently, he is serving as the Director of China Programs, Coordinator of Master of Science in Architecture Program, and an associate professor at School of Architecture Syracuse University, USA. In addition to his academic achievements, WANG Fei co-founded URSIDE Hotel Shanghai in collaboration with four other partners from various disciplines, which is an unconventional hotel offering developer, designer and manager services. He has curated many exhibitions, including "Unbuilt Eisenman (2025)," "9th Bi-City Beinnale of Urbanism\Architecture Longgang (2022)," and "Challenges and Opportunities: Architecture Innovation During the Pandemic (2021)." His design and research works have received numerous awards and have been exhibited worldwide. He has delivered speeches at numerous design and art institutions and is a prolific writer for various international journals. He serves as a columnist and guest editor for Time+Architecture and guest editor for Urban Flux. Some of his books include *Inter-Views: Trends of Architectural and Urbanism Institutions in North America and Europe* (2010) and *Poetics of Construction, The Discourse of Tectonics in Contemporary China* (2014).

王飞是一位建筑师、教育家、创业家和策展人。现任教于美国纽约雪城大学建筑学院，副教授，担任建筑理论硕士学位课程主任以及中国课程总监。他是上海有在酒店及有在设计的联合创始人。他的策展经历包括："埃森曼：未与建（2025）""深圳双年展龙岗分展场（2022）""机遇与挑战：雪城建筑疫情期间的教育创新（2021）"等。他获得众多国内外设计大奖，设计与研究作品曾在世界各地展出，曾在世界各地进行理论研究和建筑实践的讲座。他出版中英文学术论文百篇，并担任多个杂志的客座编辑与专栏作家，著作包括《交叉视角：欧美建筑城市院校动态访谈精选》《建构理论与当代中国》等。

YIN Yujun is the founder of Atelier Alternative Architecture (AAA) and an adjunct Assistant Professor in Hong Kong University's Architecture School. He holds a Master's degree in Architecture and a Master's degree in Landscape Architecture from Harvard University Graduate School of Design, where he received the dean's scholarship and research fund. In 2017, he was a visiting professor at the School of Architecture, Syracuse University, and the curator of the 2017 UABB (Shenzhen) Guangming Sub-venue. In 2015, he curated the young architects group exhibition at the 2015 UABB (Shenzhen) and served as the production manager of the chief curator team. He participated in UABB (Shenzhen) in 2005, 2007, 2009, and 2011, and the 2011 Venice Biennale. He also co-curated the 10 Million Units: Housing an Affordable City exhibition in 2011. In recent years, he participated in various exhibitions on behalf of PAO, including the China Architecture Exhibition in Berlin, the Beijing Design Week - Architecture China 1,000 exhibition, the UED 10×100 - An Exhibition of 100 Architects on the 10th Anniversary, UABB (Shenzhen) in 2013 and 2015, the 2016 Qianhai Public Art Season, and Taikang Space's "BUILDING ISSUES" Exhibition.

尹毓俊是 Atelier Alternative Architecture 工作室创始人，主持建筑师。哈佛大学设计学院建筑学及景观建筑学双硕士。2017 年雪城大学访问教授，现任教于香港大学建筑学院及哈工大深圳分校建筑学院。曾任 2017 及 2019 年深港建筑\城市双城双年展光明展场总策展人。他的研究和实践集中在复杂社会背景和城市环境中，建筑如何作为空间手段介入并影响城市环境，创造出积极的公共空间和丰富的城市公共生活。同时，他作为参展人参与多次国际性展览，如 2015 年深港城市\建筑双城双年展主题馆——青年建筑师集群展、柏林中国建筑展、北京设计周建筑中国 1000 展、UED 10×100 十年百名建筑师展、深港双年展、泰康空间"建筑问题"展。

Leslie LOK is a co-founder of HANNAH and an assistant professor at Cornell University Department of Architecture. HANNAH is an experimental design and research studio that specializes in a wide range of scales, from furniture to buildings and urbanism. In collaboration with Sasa Zivkovic, LOK has a strong interest in architectural explorations that involve material expression, digital fabrication, and innovative construction techniques. HANNAH's work aims to explore the relationship between machine means and architectural ends, employing digital design and fabrication technologies from the ground up to facilitate new material methods, tectonic articulations, environmental practices, technological affordances, and forms of construction.

陆唯佳是 HANNAH 的联合创始人，现任康奈尔大学建筑系助理教授。HANNAH 是一个前沿实验设计与研究事务所，项目跨越从家具到建筑和城市设计多个尺度。陆唯佳与萨沙·日夫科维奇于 2014 年共同创立了 HANNAH，专注于根植于推进建筑实践的材料表达、数字制造和创新结构科技的建筑探索。事务所的项目旨在挖掘机器方式和建筑终端之间的张力。从坯土建造开始，HANNAH 的项目最核心的是数字设计和制造技术，从根本上促进了新的建造方法、构造节点、环境的实践、技术的可供性和建造的形式。

FENG Guochuan is the Chief Architect of Zhubo Design Co., Ltd., founder of Archild Communication (Shenzhen) Co., Ltd., and Deputy Dean of YSBE College, Haikou University of Economics. Feng participates in the process of urbanization in China through the practice of architecture and urban design. At the same time, he reflects on the process of urbanization through writing, research, and education. He conducts practice and reflection in the dimension of space politics, observes the operating mechanism of ideology in the process of urbanization, and pays attention to the interaction between space and subject and the fate of the subject at present. He has been engaged in large-scale public building design and urban design for many years and has won many national, provincial, and municipal design awards. While engaged in architectural design, Feng is also skilled at interdisciplinary integration of architectural design, urban design, landscape design, lighting design, public art, and other fields to improve the overall aesthetic level of the city. In recent years, Feng has used his spare time to try his hand at architectural education for children, hoping to help children grow up happily, physically, and mentally through architecture.

冯果川是筑博设计股份有限公司及执行首席建筑师，童筑文化儿童教育创始人，海口经济学院雅和人居工程学院副院长。他通过建筑和城市设计实践参与中国城市化进程，同时也通过写作、研究、教育等方式对城市化进行反思。无论实践还是反思都有着清晰的空间政治学维度，观察意识形态在城市化过程中的运作机制，关注空间与主体的相互作用以及当下主体的命运。多年从事大型公共建筑设计、城市设计，获得多项国家、省、市级设计奖项。从事建筑设计的同时，善于跨界整合建筑设计、城市设计、景观设计、灯光设计、公共艺术等不同领域提高城市的整体艺术水准。近年业余时间尝试面向儿童的建筑教育，希望通过建筑学帮助儿童身心智和谐、快乐地成长。

ZENG Renzhen (aka Yu Shan) is an architect dedicated to the study of traditional Chinese gardens and paintings. He previously worked at 100s+1 Studio. In 2014, ZENG founded "Huan Yuan Studio," which focuses on studying the relations between Chinese gardens, Shanshui, space, and humankind through drawing. So far, he has created a number of serial works such as "Fantasy" and "Red."

曾仁臻，号"鱼山"，建筑师，长期执着于有关中国园林和绘画的研习，曾工作于百子甲壹工作室。2014 年创立幻园工作室，创作了大量有关中国园林、山水、空间与人的关系的研究性画作，包括"幻"、"红"等多个系列。已出版专著《幻园》《幻园第二辑：借天工》《草间居游》《造境记》。

253

As an architect, TANG Yu is a co-founder of Atelier Archmixing, an architecture studio along with ZHUANG Shen, REN Hao, and ZHU Jie. His works have received several awards such as the Gold of Commercial Buildings Award of AAA, the 2nd place Refurbishment in 2019 Archidaily's Architecture Award, the Interior Architecture Honor Award of AIA Shanghai | Beijing Design, and the WA Award of Chinese Architecture. His works have also been invited to be exhibited at various places such as the Biennale Architectural at Venice, "Towards A Critical Pragmatism: Contemporary Architecture in China" held at Harvard University, SUSAS, and Bi-City Beinnale of Urbanism\Architecture. As a crossover cultural figure, he is the director of Nextmixing, an organization dedicated to the crossover exchange of design, art, and technology. He has been invited to participate in the planning, design, and organization of several large-scale events such as China Brand Day, China Import and Export Fair, and SUSAS. At the same time, he serves as an advisor to Yitiao TV, an organization committed to promoting the recognition of excellent works of Chinese contemporary architecture by the public.

作为建筑师，唐煜与庄慎、任皓、朱捷合伙共同主持阿科米星建筑设计事务所。作品曾获亚洲建筑师协会商业建筑类金奖，Archadaily年度最佳改造建筑大奖、AIA上海卓越设计奖室内设计奖、WA中国建筑奖等；并受邀参展威尼斯双年展、哈佛大学当代中国建筑展、上海城市空间艺术季等等。作为文化类跨界人士，担任米行文化机构的主理人，致力于设计、艺术和科技的跨界交流，并受邀参与数届中国品牌日、中国进出口博览会、上海城市空间艺术季等大型活动的策划、设计、组织等工作。同时，作为"一条"视频的顾问，致力于推动中国当代建筑优秀作品被大众认知。

ZHANG Shuo is the Founder and Creative Director of SURE Design. SURE Design was founded in 2015 and is located in Shenzhen. The main business includes brand vision, exhibition vision, guide and environmental graphics, packaging, books, and other graphic design work. The company's works have won various professional awards both locally and internationally, including the GDC AWARD 2019 July Award and Professional Group Guide system Design Best Award, Award360° Design Award, China International Poster Biennale, Hong Kong Global Design Award, France i-ding Aiting Award, Taiwan Golden Point Design Award, CGDA International Design Award, Macau Design Biennale, and IAI Design Award.

张烁是烁设计的创始人和创意总监。烁设计成立于2015年，位于深圳。主营业务包括品牌视觉、展览视觉、导视及环境图形、包装、书籍等平面设计工作。公司作品曾获GDC AWARD 2019评审特别奖及专业组导视系统设计最佳奖、Award360°年度设计奖、中国国际海报双年展、香港环球设计大奖、法国i-ding艾鼎奖、台湾金点设计奖、CGDA国际标志设计奖、澳门设计双年奖、IAI设计奖等多个海内外专业类奖项，也获得了《APD亚太设计年鉴》《Wallpaper》《DOMUS》《Graphic Fest》等专业媒体的选登及报道。

MENG Hao is the Founder & CEO of RoboticPlus.AI, a PhD candidate in the direction of Construction Intelligence at Tongji University, and the leader of the Laboratory of Construction Robots and Artificial Intelligence at Zhejiang University, who has more than ten years of experience in the field of construction and robotics. In 2022, he was selected as one of the World Economic Forum Global "Technology Pioneers." RoboticPlus.AI is dedicated to developing intelligent industrial robot systems in the construction field and deeply develops core technologies such as control software, intelligent algorithms, and human-robot interaction for construction robots to empower the global construction intelligent manufacturing industry.

孟浩是大界机器人创始人和CEO，同济大学建筑机器人方向博士研究生，浙江大学建筑机器人与人工智能实验室技术负责人，兼任上海市数字建造工程技术中心机器人技术顾问，具备10+年建筑与机器人相关领域经验，入选2022年世界经济论坛全球"科技先锋"。孟浩于2016年成立了建筑科技公司大界机器人，是国内领先的建筑机器人产品公司。大界致力于开发建筑领域的智能化工业机器人系统，深耕建筑机器人的控制软件、智能算法与人机交互等核心技术，为全球的建筑智能制造行业赋能。

QIN Qingxia is the creative director of Zhejiang Zidao Culture Communication Co., Ltd. He graduated from the School of Architecture, Tsinghua University with a bachelor's degree in architecture and obtained a master's degree in landscape architecture from GSD, Harvard University. He used to be an Assistant to the Director of the China Region and a Project Manager at LSG Landscape Architecture Ltd. in Tysons Corner, VA, USA. After returning to China, he has been active as a manga artist, architect, and Weibo influencer known as "Shituzi". His representative works include A Dog, The Furious YAMA, Scientific Methodology of being A Superhero, Xiaoyaoyou, Brainstorm Supermarket, Monkey Housailei, etc. Among them, The Furious YAMA, A Dog, and other comics have started at the top in the Chinese manga market. He has won the title of Top Ten Most Influential Manga&Anime Influencers of Weibo for five consecutive years. As an architect, Athens Qin has developed video series, focusing on design and aesthetics, and worked with some of the most prestigious firms in the industry.

覃清硤是浙江子道文化传播有限公司创意总监。本科毕业于清华大学建筑系，后获美国哈佛大学硕士学位。曾担任美国LSG Landscape Architecture 公司景观设计中国区总监助理，归国后以漫画家、建筑师、微博大V"使徒子"的身份活跃。代表作有《一条狗》《大王不高兴》《科学超能方法论》《逍遥游》《脑洞超市》《猴腮雷》等，其中《阎王不高兴》《一条狗》等漫画成为国漫头部IP。连续5年获得微博十大最有影响力动漫大V的名号。使徒子凭借自身建筑师身份，开发全新的视频产品，聚焦设计美学科普，并和行业里的很多大咖达成合作

DING Junfeng is an associate professor at the School of Design and Innovation at Tongji University. He received his Master of Design in Digital Direction from Harvard University, and he is a licensed architect in the USA. In 2013, together with MIT Fablab, DING set up the first Fablab in China. He founded the Fablab O (FABO) brand, which has gained prominence in the maker society both domestically and internationally. FABO also develops STE (D) M courses and systems for colleges, vocational high schools, and K12 schools locally. Meanwhile, Professor Ding has been actively involved in interdisciplinary projects and research on digital fabrication.

丁峻峰是同济大学设计创意学院副教授，他从哈佛设计学院数字设计方向毕业并获得美国注册建筑师称号。2013 年，丁峻峰老师的主导联合 MIT 媒体实验室，在中国成立了第一家"数制"工坊实验室，并独立创办了 FABO 品牌，致力在创客创业前沿，并努力为中国大学、职业学院及基础教育设计 STE (D) M 教育课程和系统，在国内外享有盛誉。同时，丁教授也活跃在 AI 数字媒体和展览实践前沿。

WANG Xin is a Professor at the School of Architecture at China Academy of Art, the founder of Gardening Studio, and the Editor-in-Chief of Wuyou Yuan (Arcadia). He has been involved in the research and teaching of traditional Chinese literati Garden Art and is dedicated to exploring and innovating contemporary Chinese local architectural design. His notable works include "Little Cave Sky", "Tiger Art", "The Tang and Song Dynasties Next Door," and more. He has published works such as *An Architecture Towards Shanshui, A Method From Shanshui*, and has edited the series *Arcadia*. Garden Architecture was founded in 2015 and adheres to the parallel interaction between teaching and practice, transforming the results of teaching and research into social practice in contemporary China.

王欣是中国美术学院建筑艺术学院教授，造园建筑主持建筑师，《乌有园》主编。他从事传统中国文人造园艺术的研究与教学，致力于当代中国本土建筑设计的探索与创新。代表作品有：《小洞天》《虎美术》《隔壁唐宋》等。著作《如画观法》《模山范水》，编著《乌有园》系列。造园建筑创立于 2015 年，造园建筑坚持教学与实践的双线平行互动，将教学研究成果转化为当代中国的社会实践，内容涵盖建筑、园林与展览空间的营造以及道具与器物的研发。以造园为名，即是以中国园林的当代实验为己任，以山水园林的角度重建中国本土建筑学为理想。

WANG Zigeng is the founder and principal architect of PILLS, an Associate Professor, and Deputy Head of the Department of Architecture, and Head of the Media and Exhibition Center at the Central Academy of Fine Arts. He was also a Visiting Professor at the School of Architecture at Syracuse University and Tsinghua University. WANG obtained his Master of Architecture at Princeton University and is a Ph.D. Candidate at College of Architecture and Urban Planning, Tongji University. He is also a member of the American Institute of Architects and was a lecturer at the Beijing Film Academy School. His fields of research include Environmental Technology, Architecture Exhibition, Architectural Media, and Narration.

王子耕是 PILLS 建筑工作室主持建筑师。中央美术学院建筑学院副教授、建筑系副系主任、媒体展览部主任。普林斯顿大学建筑学硕士。同济大学建筑学博士候选人。美国建筑师协会会员。曾任雪城大学客座教授、北京电影学院美术学院讲师。研究领域包括环境技术、建筑展览、建筑媒介与叙事。

JU Bin founded Horizontal Design in 2003 and has gained considerable recognition for his architecture and design projects. He was named one of the ten most important emerging designers in China in 2015 and was selected as GQ Designer of the Year in 2017. Ju Bin has been recognized numerous times as the Most Influential Designer in China, and in 2020, he was named Tatler Cultural Figure of the Year. His work has been extensively published and exhibited at the Venice Biennale, Milan Design Week, and the Shenzhen/Hong Kong Urbanism and Architecture Bi-City Biennale. Ju Bin also teaches as a visiting critic at Syracuse University, was a visiting critic at the Politecnico di Milano, and serves as a graduate advisor at the Sichuan Academy of Fine Arts. Ju Bin is also the founder and council member of the C-Foundation, a public nonprofit organization committed to promoting design education.

琚宾是深圳水平线设计的创始人。该公司成立于 2003 年，从事建筑、室内、景观和产品设计。他致力于研究中国文化在建筑空间中的应用与创新，通过个性化和独特的视觉语言表达设计理念，用新的视觉传达诠释中国文化元素。在设计作品中，琚宾将"当代性"、"文化主义"和"艺术性"的语言融入设计中。他一直从过去与现在的联系与冲突中寻求设计的突破，从艺术与生活的交集中探索设计的本质，通过历史与当代思想的结合寻找设计文化的精神诉求。水平线设计是中国当代设计的代表之一，是拥有多名优秀的年轻设计师的国际化团队。琚宾现担任雪城大学建筑学院访问教授，曾任教于米兰理工大学和四川美术学院。他也是推动设计教育的非盈利组织创基金的创始成员。

Onesight Technology was founded in January 2018 by LUO Feng, an MBA graduate from Fudan University and a Bachelor of Architecture from Zhejiang University, with 10 years of experience in the architectural design industry. The company is dedicated to architectural technology and innovation. The co-founder, JIANG Tong, holds a Bachelor's degree in Automation from Zhejiang University and a Bachelor of Honors from Zhu Kezhen College. Onesight Technology offers BIM data visualization management and intelligent services for the entire life cycle of buildings. It integrates BIM into the real scene through AR/MR, AI, and other technologies to assist in the display of design results, construction process management, and operation and maintenance management. The company has built a rich and innovative product matrix, accumulated profound advantages in AR/MR technology, AI technology, IoT technology, and other fields, and has a high level of BIM data production and optimization capabilities. Onesight Technology provides high-quality BIM consulting, construction process management software, BIM operation, and maintenance visualization software, and other products and services for real estate development, engineering management, and other fields.

以见科技，成立于 2018 年 1 月，创始人为复旦大学 MBA、浙江大学建筑学学士罗锋先生，拥有 10 年建筑设计行业经验，致力于建筑科技与创新；联合创始人为浙江大学自动化学士、竺可桢学院荣誉学士蒋童先生。以见科技为建筑全生命周期提供 BIM 数据可视化管理与智能服务，通过 AR/MR、AI 等技术将 BIM 融入实景，辅助设计成果展示、施工过程管理以及运营维护管理。公司构建了丰富和创新的产品矩阵，在 AR/MR 技术、AI 技术、IoT 技术等领域积累了深厚的优势，并拥有高水平的 BIM 数据生产与优化能力，为地产开发、工程管理等领域提供高质量的 BIM 咨询、施工过程管理软件、BIM 运维可视化软件等产品与服务。

FUN was founded in 2008 by WANG Zhenfei and WANG Luming as a design and research studio consisting of architects, designers, and programmers. They strive to produce unexpected design results by incorporating knowledge from diverse fields. FUN cooperates with people from various professions, such as artists, fashion designers, mathematicians, and engineers, to explore new design possibilities through cross-disciplinary cooperation. Both WANG Zhenfei and WANG Luming earned a B.Eng. from Tianjin University and an M.Arch. from Berlage Institute.

FUN 是一个由建筑师、设计师、程序设计师组成的致力于设计和研究的事务所，由王振飞和王鹿鸣于 2008 年创立。他们致力于将不同领域的知识创造性地带入设计，希望基由不同的设计方法和设计过程创造出"意料之外"的结果。事务所与包括艺术家、服装设计师、数学家、工程师在内的各界人士进行跨界合作，并希望通过这样的合作探索设计新的可能性。他们均毕业于天津大学获得工程学学士，以及荷兰贝尔拉格学院获得建筑学硕士。

Established in 2013, Drawing Architecture Studio (DAS) explores the possibilities of drawing, space, and urban studies in a unique way. They use engineering drawing software as a tool and draw inspiration from architecture, art, popular culture, and daily life to create magnificent and complex images for the urban landscape. At the same time, they consider infinite vector images as building materials, explore the multiple paths to bringing virtual images back to the material world, and break the boundary between image and space. DAS understands urban studies as an experience and narrative with open forms, and conducts experiments through various media such as architectural models, art installations, comics, and books. LI Han, founding partner, is a registered architect in China who graduated from the Central Academy of Fine Arts and RMIT University. HU Yan, founding partner, graduated from Concordia University.

绘造社于 2013 年在北京创立，以独特的方式探索绘画、空间和城市研究的可能性。他们以工程制图软件为工具，从建筑、艺术、流行文化、日常生活中吸取灵感，创作宏大复杂的城市景观图像。同时，他们将无限大的矢量图像视为建筑材料，探索虚拟图像回归物质世界的多重路径，打破图像与空间的边界。绘造社将城市研究理解为形式开放的体验和叙事，通过建筑模型、艺术装置、漫画、书籍等多种媒介进行实验。绘造社创始合伙人李涵为国家一级注册建筑师，毕业于中央美术学院和皇家墨尔本理工大学。创始合伙人胡妍毕业于康考迪亚大学。

（设计师排序不分先后）